T0185720

SpringerBriefs in Applied Sciences and Technology

SpringerBriefs present concise summaries of cutting-edge research and practical applications across a wide spectrum of fields. Featuring compact volumes of 50 to 125 pages, the series covers a range of content from professional to academic.

Typical publications can be:

- A timely report of state-of-the art methods
- An introduction to or a manual for the application of mathematical or computer techniques
- A bridge between new research results, as published in journal articles
- A snapshot of a hot or emerging topic
- An in-depth case study
- A presentation of core concepts that students must understand in order to make independent contributions

SpringerBriefs are characterized by fast, global electronic dissemination, standard publishing contracts, standardized manuscript preparation and formatting guidelines, and expedited production schedules.

On the one hand, **SpringerBriefs in Applied Sciences and Technology** are devoted to the publication of fundamentals and applications within the different classical engineering disciplines as well as in interdisciplinary fields that recently emerged between these areas. On the other hand, as the boundary separating fundamental research and applied technology is more and more dissolving, this series is particularly open to trans-disciplinary topics between fundamental science and engineering.

Indexed by EI-Compendex, SCOPUS and Springerlink.

More information about this series at http://www.springer.com/series/8884

Santiago García-Yuste

Sustainable and Environmentally Friendly Dairy Farms

 Springer

Santiago García-Yuste
Faculty of Chemical Sciences
and Technology
University of Castilla-La Mancha
Ciudad Real, Spain

ISSN 2191-530X   ISSN 2191-5318   (electronic)
SpringerBriefs in Applied Sciences and Technology
ISBN 978-3-030-46059-4   ISBN 978-3-030-46060-0   (eBook)
https://doi.org/10.1007/978-3-030-46060-0

This Springer imprint is published by the registered company Springer Nature Switzerland AG
The registered company address is: Gewerbestrasse 11, 6330 Cham, Switzerland

Introduction

Rapid climate change is driving global temperature rise, warming oceans, shrinking ice sheets, retreating glaciers, decreasing snow cover, rising sea levels, declining artic sea ice, acidification of oceans and extreme weather events. The Intergovernmental Panel on Climate Change stated that "Scientific evidence for warming of the climate system is unequivocal".

The current warming trend is threatening us, and we must find the tools to combat it as soon as possible. We do have to be concerned about the emission of any GHGs on the Earth. CO_2 is the greenhouse gas that traps most heat, and its emissions have increased since the industrial revolutions. CO_2 is released through burning fossil fuel, deforestation, volcanic eruptions and anthropogenic activities. Every day, international institutions such as the UN, FAO, NASA, USEPA, USDA or IEA warn us about the need to transform anthropogenic actions to more sustainable processes, whatever the activity.

Data reveal that more than half of all CO_2 emissions originate from small and dispersed sources, such as agricultural operations or the transportation sector. Husbandry processes are the dispersed source of greatest concern to the international community. In fact, an estimate of c. 6,500.0 kg of CO_2 per dairy cow per year is reported. Any novel carbon capture strategy to mitigate CO_2 emissions from relatively minor sources should be warmly welcomed.

The authors of the book consider that the best way to make people aware of the hazardousness of emissions from husbandry requires an understanding of the structure of the CO_2 molecule, the different strategies proposed to remove it and everyday husbandry activities. This book is intended to change the way people think about emissions from relatively dispersed and minor sources.

The book is divided into five interrelated chapters. In Chap. 1, the authors describe the current situation of the global dairy livestock industry, according to the Paris Agreement, and analyse the potential of the current environmental solutions and bioeconomic alternatives. Chapter 2 is discussing physiology and how the cow's metabolism produces CO_2. In Chap. 3, the authors describe the production systems of dairy farms and the management and treatment processes of the waste they generate, highlighting the most commonly used means of mitigating environmental impact. The aim of Chap. 4 is to update the readership on the different strategies that have been developed, are in development or are merely proposed for dealing with the reduction of CO_2 from the atmospheric pool. Finally, Chap. 5 describes the 'CO$_2$-RFP Strategy', as an innovative carbon dioxide use proposal for the dairy industry. The predictable CO_2 and NH_3 emissions from the dairy cow in a barn, can be converted into ammonium bicarbonate, one of the most widely used fertilizers in the world. The implementation of the 'CO$_2$-RFP Strategy' will transform a conventional dairy farm into a 'sustainable and environmentally friendly farm'.

Dr. Santiago García-Yuste

Contents

1 The Sustainability Challenge of Dairy Livestock Systems 1
Jorge Zafrilla, Ángela García-Alaminos, and Fabio Monsalve
1.1 Introduction 1
1.2 Dairy Sector Within the Sustainable Development
Goals and Climate Change Framework 3
1.3 Global Dairy Activity, Greenhouse Gas Emissions
and Carbon Footprint 6
1.4 Environmental Solutions and the Bioeconomy 12
1.5 Conclusions 14
References 15

**2 The Ruminant: Life History and Digestive Physiology
of a Symbiotic Animal** 19
Francisco Javier Pérez-Barbería
2.1 Taxonomy, Evolution and Feeding Styles
of the Ruminant Animal 20
2.2 Domestication of the Ruminant Animal 23
2.3 Why the Ruminant Is Different: Adaptations to Diet 24
2.3.1 Mouth, Teeth, Body Size and Salivary Glands 24
2.3.2 The Ruminant Stomach 24
2.3.3 Small Intestine 27
2.3.4 Caecum and Large Intestine 29
2.4 Cattle: A Large Contributor to Greenhouse Gas (GHG)
Emissions 29
2.4.1 Cattle Taxonomy, Habitat and Diet 29
2.5 Ruminant: The Efficient Animal 30
2.6 Use of Structural and Non-structural Plant Cell Components 33
2.7 Carbohydrate Metabolism 35
2.8 Protein Metabolism 37

    2.9   Lipid Metabolism   39
    2.10 Conclusion   41
    References   42

3   Husbandry: Milk Production   47
    Abdessamad Gueddari and Jesús Canales Vázquez
    3.1   Introduction to the Dairy Production Systems   47
           3.1.1   Extensive Production Systems   49
           3.1.2   Intensive Production Systems   50
           3.1.3   Mixed Production Systems   51
    3.2   Milking Systems   51
           3.2.1   Tied-Stall Milking Systems   52
           3.2.2   Parlour Milking   53
           3.2.3   Automated Milking Systems   57
    3.3   Farm Construction Elements   58
           3.3.1   Enclosures and Roofs   58
           3.3.2   Flooring Considerations   59
    3.4   Environmental Control   60
           3.4.1   Heat Stress   60
           3.4.2   Ventilation and Air Quality   61
           3.4.3   Heating and Cooling   63
    3.5   Waste Management   64
           3.5.1   Manure Management   65
           3.5.2   Anaerobic Treatment   72
           3.5.3   Future Prospects   74
    3.6   Conclusion   77
    References   77

4   Quick Fire Set of Questions About CO_2 that Need
    to Be Answered   81
    Carlos Alonso-Moreno and Santiago García-Yuste
    4.1   Is the Molecule of CO_2 Dangerous?   81
    4.2   So, Should I Be Worried About the CO_2 Concentration
          in the Atmosphere?   81
    4.3   What Is the Carbon Cycle?   82
    4.4   Can We Do Something to Balance the Global Carbon
          Budget Again?   84
    4.5   Are There Any Technologies Able to Capture the CO_2
          Directly from the Air?   85
    4.6   What About Making Useful Products from CO_2?   86
    4.7   Should We Forget About CDU Strategies?   86
    4.8   What Are the Most Important CDU Strategies?   87
    4.9   Can We Use the CO_2 After Storage?   88

4.10 Can You Explain a Little Bit More About the CO_2-NH_3
    System? 88
4.11 Are There Any CDU Strategies Proposed for Minor
    Sources? 89
4.12 What Is the Meaning of NETs? 91
4.13 How Can We Analyse the Sustainability of the Different
    Strategies Proposed? 94
References 96

5 The 'CO$_2$-RFP Strategy' 99
Carlos Alonso-Moreno and Santiago García-Yuste
5.1 Introduction 99
5.2 CO_2 and NH_3 Estimated Emissions from Intensive Husbandry
    Production Systems...................................... 102
    5.2.1 Regarding CO_2 Emissions 102
    5.2.2 Regarding NH_3 Emissions 104
5.3 The *CO$_2$-RFP Strategy* 105
    5.3.1 The *CO$_2$-RFP Strategy* as a Business Model 106
    5.3.2 The *CO$_2$-RFP Strategy* Regarding Negative
        Emissions 108
    5.3.3 The *CO$_2$-RFP Strategy* with Regard
        to Sustainability 109
5.4 Conclusions 110
References 110

Chapter 1
The Sustainability Challenge of Dairy Livestock Systems

Jorge Zafrilla, Ángela García-Alaminos, and Fabio Monsalve

Abstract In this chapter, the authors describe the current situation of the global dairy livestock industry under the influence and challenge of the commitments of the so-called Paris Agreement. Firstly, the key points of the Agreement affecting the livestock and dairy systems are discussed within the framework of the Sustainable Development Goals. Next, a detailed analysis of the evolution of the activity and the greenhouse gas emissions of the industry is presented. Finally, a summary of current environmental solutions and bioeconomy alternatives will contribute to enriching the discussion.

1.1 Introduction

Recent estimates from the United Nations (UN) expect an increase from 7.6 billion people today to 9.8 billion people by 2050 (UN 2019). Together with the population increase, food demand is expected to more than double by 2050 because of increases in living standards (Rojas-Downing et al. 2017). Agricultural systems are called on to produce the extra food required to secure the food supply of millions of people. Considering that one of the bases of the current global diet is consumption of animal products, especially liquid or processed milk products, which are consumed by more than 6 billion people, the challenge for the dairy sector is very clear (FAO and GDP 2018).

However, the current state of climate crisis jeopardises the ability of agricultural systems to provide this additional new and secure supply of food within the standards set by Sustainable Development Goals (UN 2015b). The role played by the agricultural sector in the coming decades will be crucial, as at the same time they will have to deal with goals related to mitigation of and adaptation to climate change. Agricultural systems find themselves in a situation where, on the one hand, it is one of the industries at the greatest risk of facing extreme climate events that may endanger the basic livelihoods of hundreds of millions of people; and, on the other hand, agricultural activities are considered to be one of the main drivers of global greenhouse gas

S. García-Yuste, *Sustainable and Environmentally Friendly Dairy Farms*,
SpringerBriefs in Applied Sciences and Technology,
https://doi.org/10.1007/978-3-030-46060-0_1

emissions (GHG). Although there is no international agency reporting official figures about the GHG of agriculture, forestry and other land use (AFOLU), in the way that there do exist precise figures provided by, for example, the International Energy Agency (IEA) for fossil fuel combustion-related emissions (Tubiello et al. 2013), some studies set global agricultural emissions at 10–12% according to Reisinger and Clark (2018) or 15% according to Persson et al. (2015), with 80% of total emissions due to ruminants. To these emissions should be added emissions from deforestation, energy use and animal-feed production, so that the total percentage of emissions from agricultural activity seems to be key given the current situation.

A deeper depiction of the emissions of agricultural systems is needed, as this represents the highest non-CO_2 emissions source. Agriculture accounts for more than 40% of global methane (CH_4) emissions, and 60% of nitrous oxide (N_2O). These emissions come from the increasing use of synthetic fertilizers, manure use and enteric fermentation. However, agricultural production increased by 70% over the period 1990–2015, which indicates a significant improvement in GHG efficiency (Frank et al. 2019). Also, the increase in agricultural emissions is taking place at a slower rate than fossil fuel emissions, which results in a reduction of the share of agriculture's contribution to GHG global emissions (Tubiello et al. 2013). It is important to highlight that notable differences arise when the source of emissions, or the region where they are generated, is analysed. Emissions derived from the use of fertilizers have grown more than from the use of manure. By regions, emissions in developing countries have grown more than in developed regions, and they have even fallen in wealthy regions, as is the case of Europe (Tubiello et al. 2013).

Most of the emissions from agriculture come from livestock, mainly ruminants, and action in this sub-sector is thus key to achieving climate objectives (Reisinger and Clark 2018). In the same way as the whole agriculture sector, GHG livestock inventories have significant measurement problems. In most cases, emissions estimates use global warming potentials (GWPs), which may lead us to underestimate the effects on climate of other non-CO_2 gases such as CH_4 from enteric fermentation and manure management, and N_2O from manure management and nitrogen deposition (Reisinger and Clark 2018). The case of CH_4 is clear; it does not accumulate and behaves in the same way as CO_2 in the atmosphere. Its life cycle is shorter, but the warming potential is much higher per molecule and kilogram. Other studies such as Persson et al. (2015) also identify problems with the use of these GWP measures, as they can return poor estimates.

The Paris Agreement (UN 2015a), the most ambitious global agreement on the fight against climate change, sets clear targets for GHG emissions to achieve the goal of at most a 2 °C rise by 2100. Among other goals, it aims at net-zero CO_2 emissions derived from anthropogenic action, mainly through the decarbonisation of energy, which seems a relatively plausible medium-term goal (Peters et al. 2017). This may cause the share of other non-CO_2 gases in the GHG total to increase decisively towards the 2100 scenarios. Action to reduce emissions will have to be encompassed in parallel with dealing with the special characteristics of livestock systems as they play a key role in meeting the commitments of the Sustainable Development Goals

(SDGs), such as ensuring food security and other areas related to draught power, nutrient cycling, social and capital insurance, and so on (Reisinger and Clark 2018).

Countries need to have good GHG inventories for livestock in order to be able to properly identify the objectives of their nationally determined commitments (NDCs), and to be transparent. Given the difficulties and challenges posed by the livestock industry, especially in developing countries, direct emissions measurement systems, including measures such as productivity with more advanced inventories, are essential for success in the fulfilment of climate and sustainable development objectives (GRA and CCAFS 2016). The productivity measures and the control of energy intake that the animal needs will return better estimates of GHG that will help countries to provide better measures of emissions, and will not jeopardise food security, since the levels of milk or meat production in animals can be maintained or even increased; thus the goal of reducing GHG emissions will be attained by improving efficiency and productivity (GRA and CCAFS 2016).

In the dairy sector, an 18% increase in emissions was observed between 2005 and 2015, resulting from a 30% overall increase in production, motivated by the rise in final demand. These figures show an improvement in efficiency resulting from the increase in production that is far greater than the increase in emissions. Improvements in efficiency have helped to avoid reaching an expected increase of 38% of total emissions, causing the intensity per unit of output to fall (FAO and GDP 2018). These data are found qualitatively in all regions, although the improvement in intensity in developing regions is less than in developed regions. In developed regions emissions range from 1.3 to 1.4 kg CO_2-eq., kg fat-and-protein corrected milk; while in developing regions, the range is from 4.1 to 6.7 kg CO_2-eq., per kg fat-and-protein corrected milk in 2015. Management practices used in these countries are behind these differentials, so there is room for improvement in the medium term to further reduce emissions (FAO and GDP 2018).

1.2 Dairy Sector Within the Sustainable Development Goals and Climate Change Framework

Sustainable Development Goals. The dairy sector emerges as a key industry in ensuring commitment to most of the global Sustainable Development Goals (SDGs) listed by the United Nations (UN 2015b). The Global Dairy Agenda for Action (GDAA 2015) identifies a list of key elements that differentiate the dairy sector from others in terms of sustainability: Currently, about 1 billion people make their living in or around the dairy sector; billions of people enjoy the nutritional benefits of consuming milk and dairy products; rural communities created around the sector improve quality of life and reduce social inequality, alleviating poverty and unemployment; and measures such as yield improvements, feed efficiencies, the use of by-products for human consumption and manure as fertilizer are the basis of an economic, socially and environmentally friendly model.

From the institutional sphere, a number of initiatives have appeared to show this alignment of the dairy sector with sustainability. One of these is the Dairy Sustainability Framework (DSF), launched in 2013, which is a pre-competitive and globally recognised sustainability initiative which helps the industry to improve its sustainability under 11 criteria related to GHG emissions, animal care and market development (Rabobank 2016). The DSF is a holistic approach to sustainability which is completely aligned with SDG's, with clear links and synergies between them. The dairy value chain participants in the DSF initiative benefit from it because it takes into account regional and local issues connected at the global level, promotes the alignment of the sector's actions regardless of the level, size or location, fosters the sharing of best practice, establishes a common language to monitor and share the progress and improvements, and provides a reference for in-depth communication on the actions and results obtained (DSF 2019).

From a broader perspective, the bioeconomy strategy is probably one of the most important and ambitious initiatives to integrate the livestock industry into the correct sustainability path. 'Bioeconomy' is understood as 'an economy using biological resources from the land and sea, as well as waste, including food waste, as inputs to industry and energy production. It also covers the use of bio-based processes by green industries, those parts of the economy that use renewable biological resources from land and sea—such as crops, forests, fish, animals, and micro-organisms—to produce food, materials and energy' (European Commission 2015). The potential of the livestock sector as a key piece of a circular bioeconomic model is well-known, since this industry can generate multiple benefits, such as regulating the ecological cycles by recycling biomass, and using manure as a bio-resource, fostering landscape and biodiversity preservation, generating employment and innovation in rural areas or producing protein-rich and safe food, among others (Peyraud 2016).

Paris Agreement. Livestock industries play a crucial role in addressing the challenge of climate change. The biggest global agreement to fight the current climate crisis, the so-called Paris Agreement, achieved great consensus with 197 signatory countries (UN 2015a). To date, 185 countries have ratified the agreement to keep the increase in global average temperature well below 2 °C by the end of the current century. Each country's efforts have to be described and designed within each country's nationally determined commitments (NDCs). The goal of reducing anthropogenic emissions through the decarbonisation of the energy systems is only one part of the list of policies recommended. The Paris Agreement also considers potential strategies for negative emissions that involve the removal of CO_2 from the atmosphere, either through technologies such as afforestation and agricultural practices, or through direct CO_2 removal from the atmosphere using chemical innovations (Anderson and Peters 2016; Peters and Geden 2017; Williamson 2016). Livestock systems possess the perfect characteristics to promote actions along all the lines set out in the Paris Agreement.

In current state of NDC development, 92 countries around the world have included the livestock sectors within their future strategies for reducing environmental pressures. Developing countries have the opportunity to take the lead in implementing their NDCs as climate action in agricultural sectors. The strategies could result in positive outcomes for the fight against climate change, but could also be a key driver to achieve other SDGs related to the eradication of poverty, hunger and malnutrition (FAO 2016). This task must be undertaken with the help of the international community, as the Paris Agreement incentivises international cooperation between signatory countries to ensure the success of the implementation of climate actions and policies. All international machinery, in terms of financial institutions, R&D organisations and agencies and any other actors should enhance the commitment of the goals in developing countries and open up efficient mechanisms to give transparency to these processes, implement coherent policies, provide research tools, generate enough capacity to implement the actions and guarantee enough investment flows (FAO 2016).

The challenge does not only affect low-income, developing countries; climate change is also a threat in areas such as Europe and Central Asia, because of issues related to food security, nutrition and ecosystem services (FAO 2018). Extreme climate events are endangering the economic stability of those areas, mostly in middle-income regions. The identification of proper commitments, policy and finance gaps, and opportunities for enhancing climate change mitigation strategies is also crucial at a global level. Most of the European and Central Asian countries have ratified the Paris Agreement and have submitted NDCs together with SDG targets. Implementation of these actions will take place in the coming years. The success of these actions will depend on the quality of national planning tools, together with institutional coordination and correct technological improvements to ensure technical capacity within sectors. Developed countries within the EU-28 are setting adequate legislative and institutional frameworks to ensure commitment to the goals. However, the development of regulations in middle-income countries is still in process (FAO 2018).

Within this framework, agriculture in general, and, in particular, livestock systems, need a proper way to address their national GHG inventories, in order to improve the quality of their NDC commitments. However, many countries lack relevant data on the way they compute GHG inventories for their agriculture sectors (Wilkes and van Dijk 2018). Indicators based on the inclusion of productivity and efficiency promoted in the so-called Tier 2 approaches to estimating livestock emissions in the inventories are the keys to success in achieving goals such as the Global Research Alliance on Agricultural Greenhouse Gases (GRA) (GRA and CCAFS 2016). Wilkes and van Dijk (2018) report a complete collection of livestock Tier 2 inventory practices by countries. Networks such as the GRA facilitate links across organisations, research institutes and governments to share information, knowledge and inventories in order to improve the results of each country's NDC implementation (GRA and CCAFS 2016).

1.3 Global Dairy Activity, Greenhouse Gas Emissions and Carbon Footprint

Global dairy activity. Global figures show the production of almost 669,000,000 tonnes in 2017 of fresh cow's milk. According to FAO and GDP (2018) and FAOSTAT (2019b), raw cow milk production experienced an average growth of 2.8% per year in the period 2005–2017, which has accelerated in the last seven years. Figure 1.1 identifies the main raw cow's milk producers in 2017. The main actors are not only the emerging countries with high populations like India, Brazil or China, but also developed nations such as the USA, Germany, France or New Zealand. The European Union and NAFTA (Mexico, USA and Canada) account for almost 40% of worldwide production. Africa is the major region with lower production, together with some countries in South-East Asia and South America. The left-hand panel in Fig. 1.1 shows that worldwide production has increased since 1995, especially in emerging regions like India, China or the Rest of the World (RoW). As FAO and GDP (2018) state, the most growth observed is concentrated in regions such as West Asia and Northern Asia, South Asia and sub-Saharan Africa with 4.5, 4.0 and 3.6% growth rates per year, respectively. Whereas in the two major developed areas, NAFTA and EU, generally considered as traditional dairy cattle regions, production growth rates are below the global average, 1.5 and 1.6%, respectively, while the global average over the period 2005–2015 is about 2.8% (FAO and GDP 2018; FAOSTAT 2019b).

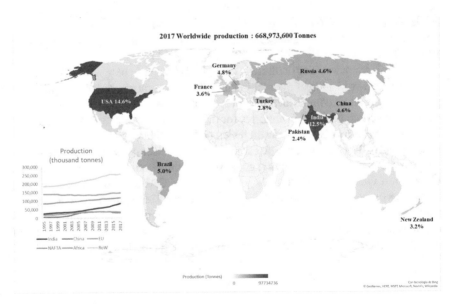

Fig. 1.1 Production of whole fresh cow's milk (tonnes), 2017 *Source* Prepared by the authors using FAO livestock primary datasets (FAOSTAT 2019b)

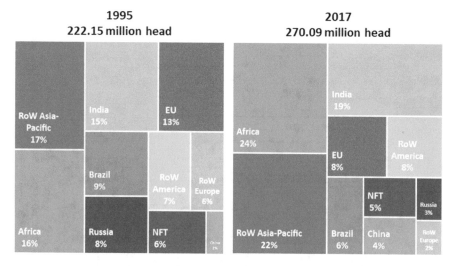

Fig. 1.2 Animals producing whole fresh cow's milk (head). Evolution in major regions in 1995 (left panel) and 2017 (right panel)[1] *Source* Prepared by the authors using FAO livestock primary datasets (FAOSTAT 2019b)

The production shares of both developed regions have declined because of the boost in production in emerging countries.

The measurement of productivity in the dairy industry is generally considered to be milk production per cow per year, which gives a measure of the yield. Figure 1.2 shows that the common trend in the period 1995–2017 is to reduce the number of cows producing fresh milk in developed regions such as Europe (both the European Union and the other European countries), NAFTA and Russia, while increasing it in emerging regions like Africa, India, China, RoW America, and RoW Asia-Pacific.

On the one hand, the reduction in developed regions can be justified by the high values of the average yield achieved by these areas in 2017 with respect to 1995: as Fig. 1.3 shows, these regions exhibit yields above the worldwide average in 2017. On the other hand, emerging regions have increased not only the number of cows producing milk but also their average yields (see Fig. 1.3). Therefore, these patterns could be indicating that developed areas are following an increasing trend of importing raw cow's milk from developing regions as a primary input to be introduced into their production chains.

Looking more closely at Fig. 1.3, the case of China is very notable. Its yield used to be below the world average, but from the mid-2000 onwards it surpassed this mean value. An increasing demand for dairy products in the Chinese market has motivated the rise of domestic milk output in the period analysed (shown in the left panel of

---

[1]For the sake of visual clarity, Figs. 1.2, 1.3 and 1.4 show a regional aggregation proposal: EU all EU-28 countries; NAFTA includes USA, Mexico and Canada; RoW Europe all non-EU-28 countries; RoW America all Central and South-American countries, with the exception of Mexico; RoW Asia-Pacific all Asian and Pacific countries with the exception of India, China and Russia.

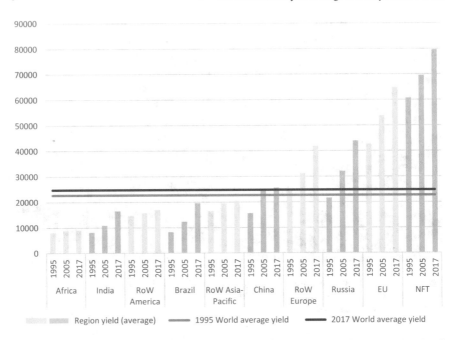

Fig. 1.3 The average yield of animals producing whole fresh cow's milk, hectograms per head. Evolution of major regions 1995–2017 and worldwide average *Source* Prepared by the authors using FAO livestock primary datasets (FAOSTAT 2019b)

Fig. 1.1), which has been sustained not only by the growth of the dairy herd shown in Fig. 1.2, but also by the significant upturns in productivity seen in Fig. 1.3.

Conversely, Fig. 1.3 shows that regions like Africa, RoW America or RoW Asia-Pacific have barely increased their yields, which are below the world average over the whole period considered. Smallholder dairy production systems are common in developing areas like Asia, sub-Saharan Africa and Latin America, but these small exploitations suffer constraints to production related to access to technological progress (for instance, the choice of species, new feed resources, improved breeding and animal health care and so on), which causes difficulties in making improvements in the production yield (Devendra 2001). Brazil had, in 1995, a similar yield to Africa or India, and much lower than RoW America's yield, but its increase has been much faster.

The productivity differentials by region are explained by the different industry structures among regions, the access to new technological developments and the

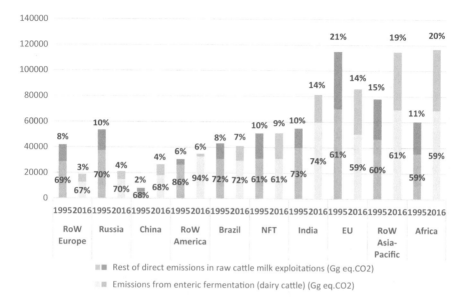

Fig. 1.4 Direct emissions from whole fresh cow's milk production: emissions from enteric fermentation and other direct emissions. Evolution of major regions 1995–2016[2] *Source* Prepared by the authors using FAO livestock primary and enteric fermentation datasets (FAOSTAT 2019b) and (FAOSTAT 2019a), respectively

unique characteristics of each region that modulate the structure of the dairy systems (FAO and GDP 2018).

The dairy greenhouse gas emissions. Dairy systems manifest increasing efficiency as emissions per unit of output are declining (Frank et al. 2019). However, the remarkable increase in milk production around the world has increased the global numbers in terms of emissions derived from dairy activities (Rojas-Downing et al. 2017; Frank et al. 2019; FAO and GDP 2018). The evolution of global emissions in the dairy sector shows an increase of about 18%, 256 million tonnes CO_2-eq. in 2015 relative to 2005 levels according to FAO and GDP (2018) figures. The amount of emissions generated differs by region and type of source considered. Figure 1.4 shows direct emissions generated within the dairy milk sector, which are disaggregated into emissions from enteric fermentation (methane gas produced in the digestive systems of the cattle) and other direct emissions. For most countries, emissions generated directly by the animal account for more than 60% of the total, and in some regions

---

[2]The lower percentage represents the participation of emissions from enteric fermentation over total direct emissions of each region. The upper percentage represents the participation of each region's total direct emissions over worldwide raw milk industry direct emissions.

where traditional production methods predominate this is even higher (in RoW America, emissions from enteric fermentation accounted for 94% of total direct emissions in 2016). This means that, although applying sustainable practices to farming and exploitation methods is always a positive action, the most effective way to reduce direct emissions from cow's milk production is to reduce the herd by increasing the yield since the bulk of emissions comes directly from the cattle's digestive system. Looking at the different regions, it is outstanding that Europe (both the EU and other countries), Russia and Brazil are the only regions that have reduced the absolute and relative values of their direct emissions in 2017 with respect to 1995. This fall can be explained by looking at the reduction in the number of animals in these regions set out in Fig. 1.2. On the other hand, RoW Asia-Pacific and Africa have moved from 15 to 11%, respectively, in total emissions in 1995 to a participation of 19 and 20% in 2016. Again, China's case is remarkable as it is the region that has most increased its direct emissions.

The continuous gains in productivity have limited the rise in emissions intensities. As stated by Gerber et al. (2011), productivity increases constitute not only a way of satisfying the increasing demand for milk but also a suitable mitigation method, especially in regions with milk yields below 2000 kg/cow/year. According to FAO and GDP (2018) measures, the emissions intensity of milk decreased by 11% over the ten years period considered, 2005–2015, from 2.8 to 2.5 kg CO_2-eq. Improvements in grassland management and feeding practices, together with improved animal genetics management, are behind these results. As FAO and GDP (2018) state, higher milk yields transform the cow's metabolism in favour of milk production and reproduction rather than maintenance. The energy intake of a high-producing dairy cow for milk production is greater than for maintenance (47%); in the case of a low-producing dairy cows the intake of energy for maintenance (75%) is greater than for milk production, which contributes, in the first case, to lowering the emissions intensity per animal (FAO and GDP 2018). More specifically, methane—which constitutes the bulk of direct emissions generated within the sector—and nitrous oxide emissions fall with increasing productivity while carbon dioxide emissions increase but to a lower extent (Gerber et al. 2011).

The global dairy carbon footprint. Common analyses accounting for dairy systems emissions use the producer criterion by region to estimate the contribution of the sector to the generation of GHG emissions. Current developments in the allocation of responsibilities change the perspective of the analysis by focusing on the consumer side using measures of carbon footprint. The concept of carbon footprint is an appropriate tool in the allocation of responsibilities as it accounts for all the emissions generated in the production chain, considering not only direct emissions but also indirect emissions, both those generated domestically and those imported from other regions. In this section, using the common modelling of environmentally extended multi-regional input–output models (EE-MRIO)[3], we have estimated the

---

[3]The carbon footprint estimates given in this chapter were developed using the common environmentally extended multi-regional input–output models (EE-MRIO) developed in papers like (Hertwich and Peters 2009; Minx et al. 2009; G. Peters et al. 2016). Both the world multi-regional

carbon footprint of the global raw milk sector (mostly represented by cow's milk production) differentiating between the regions where the highest footprint is generated and also between domestic and imported emissions flows.

Figure 1.5 uses a Sankey diagram to show the flow of the total amount of carbon footprint (kt CO_2-eq.) by emissions source, the countries where the bulk of the carbon footprint is triggered by their raw milk final demand patterns and the domestic or imported origin of the footprint for the available period 1995–2011. The total carbon footprint of the raw milk sector grew by 16% over the whole period. The left panels of Fig. 1.5 show that the most significant part of the GHG footprint in the raw milk sector (around 80% for the four years considered) corresponds to methane (CH_4), which comes mainly from enteric fermentation in the digestive systems of cattle. The right panels show that the origin of this footprint is mainly domestic. Raw milk is a perishable primary input that might not be worth importing due to its high transport costs, and so, as might be expected, a vast part of the carbon footprint of the sector is generated domestically (96.4% in 2011). The remaining imported part of the global footprint shows limited growth in the period considered, mainly generated in more developed regions such as the European Union and NAFTA. Similar but smaller trends are observed for RoW Asia-Pacific and Russia.

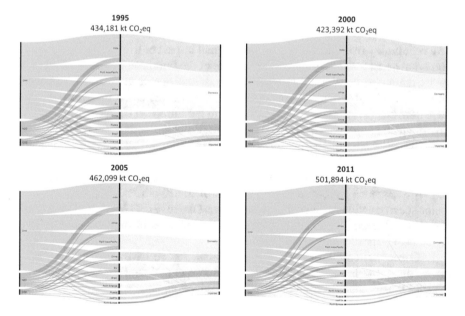

Fig. 1.5 GHG footprint of the raw milk industry in major regions. Disaggregation into GHG gases (kt CO_2-eq) and domestic/imported origin of emissions. 1995–2011 *Source* Prepared by the authors based on EXIOBASE (Stadler et al. 2018)

input–output table and the environmental satellite accounts come from the EXIOBASE. A global, detailed multi-regional environmentally extended supply-use table (MR-SUT) and input–output table (MR-IOT) (Stadler et al. 2018).

Looking at how major regions contribute to these footprints with their demand for raw milk, no big changes have occurred at the top of the list. India has by far the highest and most rapidly increasing figures in terms of carbon footprint, accounting for about 29% of the total footprint over the period. Emerging regions, like Africa, China and RoW Asia-Pacific, have expanded their footprints notably because of their increasing demand for domestic production of dairy products and, due to exports, they also supply to developed regions. Despite its vast population, China had a relatively small footprint in 1995 (7.5% of the total footprint), as it has traditionally been described as a 'lactophobe' society (Harris 1998), but it has climbed up the ranking over the period analysed, passing other regions like the European Union. This increase in the footprint of certain developing regions might be linked to the so-called 'livestock revolution' (Delgado 2003), which is primarily driven by the increasing demand for meat and milk in developing regions as they achieve higher income levels (FAO 2017). In addition, other authors point out socio-cultural changes and global advertising campaigns behind this increase in milk consumption in developing Asian regions such as India and China (Fuller et al. 2006; Wiley 2011).

1.4   Environmental Solutions and the Bioeconomy

The regulatory framework. The environmental impact of livestock and its regulation is not a new issue, especially in the sphere of the European Union. Since the late 1990s, regulatory development has constantly attempted to reduce the environmental impacts associated with the activity. As Loyon et al. (2016) stated, in the last 30 years there have been successive legislative efforts in the European Union to deal with the environmental impacts of the livestock systems. Problems derived from the presence in the water of nitrates and phosphorus have led to actions such as the implementation of the National Emissions Ceiling Directive (EC 2001, 2016)—which, since 2001, sets targets for the contention of NH_3 pollution, emissions regulations for CH_4 and N_2O livestock set out in the framework of the Kyoto Protocol (UN 1997)— sector-specific actions detailed in the Emissions Directive (EC 2010), and, of course, all actions derived from the most recent implementation of the goals of the Paris Agreement (UN 2015a).

FAO recommendations and policy actions. As we have seen in the preceding sections, the livestock sector generates more emissions than any other food-related industry, mainly from animal feed production, enteric fermentation, waste and changes in land use. The solutions to reducing the environmental impact of this sector exist, but we must act quickly as it is an intensively growing sector motivated by population growth, rising incomes and demand from developing countries. FAO considers three main lines of action to reduce the pollution intensity of the livestock sector (FAO 2017):

(1) Improvements in productivity: The central objective is the reduction of pollution intensities. FAO estimates that improved practices can contain emissions

between 20 and 30%. They recommend, on the one hand, improving the quality of animal feed and to prefer local options in doing so, and on the other, better reproductive efficiency and better veterinary care are desirable, to prolong the reproductive life of the animal. Finally, better planning of animal husbandry programs, as well as conservation of genetic diversity, is recommended to increase productivity and allow the animal to adapt to changing environments.

(2) Carbon sequestration: The maintenance of permanent pasture should be the strategy followed, given its high capacity to contain carbon. Grazing naturally plays an essential role in this. There are many options for improving the quality of pastures; for example, their efficiency can be improved by balancing, spatially and temporally, the presence of livestock, or other solutions related to fertilisation, nutrient management, and so on. However, approximately 20% of the world's grasslands have been degraded, reducing the possibility of adapting them to climate change.

(3) Integration in the circular bioeconomy: The productive model of livestock must avoid linearity; leaks of materials and energy must be redirected back to the production process. On the one hand, the correct use of crop residues and agro-industrial by-products must be guaranteed, as they represent a high percentage of the animal food base, and this is even more true at a time of expansion of world production. The use of nutrients derived from livestock manure for fertilizer production is another key to the development of a production model based on bioeconomy, because it helps producers to improve efficiency in the use of natural resources, making them more resilient to adverse climatic events. Government regulation is key in this sense, as it has to assist, encourage and certify the integration of actions that favour the bioeconomic circular model.

The bioeconomy in livestock activities. The third of these strategies is undoubtedly the most ambitious, and the one that can generate the best results in terms of sustainability, understood in a broad sense. The commitment to bioeconomy in the livestock sector in the European Union is presented in Lutzeyer (2019). Since 2005, through the 7th Framework Program, the 'Knowledge-based Bioeconomy' line has been launched, which has been followed by a multitude of meetings, programs and funding lines for bioeconomic goals. In 2012, the European Commission proposed a strategy called 'Innovating for Sustainable Growth: A Bioeconomy for Europe', which takes a bioeconomic view of the challenges facing the livestock industry, as it is expected to be increasingly affected by competition for natural resources and by the urge to reduce its environmental impact (European Commission 2012). The bioeconomy today is thus an economic sector including a multitude of activities that go beyond the agricultural sectors, such as textiles, wood and furniture, paper, the biochemical and plastic industries and energy, which by 2016 had generated almost 19 million jobs. Financing and regulatory programs currently maintain the commitment to bioeconomics in the European Union. The priorities set by the Union in this respect are job creation, mitigation of climate change, renewal and improvement of the European Union's industrial base, strengthening the circular economy based on bioeconomy and the conservation of ecosystems and biodiversity (Lutzeyer 2019).

Specifically, the European Union is focused on the livestock sector as one of the main motors of the new bioeconomic model, given that the sector is a perfect example of a circular model for the use of waste and its reincorporation into production processes (Peyraud 2016). The European Union highlights the benefits brought by this sector in terms of food security and healthy diets, given that it represents the nutritional base of millions of households. Current EU bioeconomy programs consider actions for basic livestock, with the aim of contributing to improving efficiency in the agricultural sectors, as synergies between crops and livestock are key, and even more so at a local level (Peyraud 2016). In addition, farming systems can provide public goods such as agricultural landscape, farmland biodiversity, air and water quality, food security and rural vitality, so a public good-oriented bioeconomy can bring additional opportunities to society, as it empowers small-size exploitations to make their own contribution to the SDG (Schmid et al. 2012). The European Union promotes the improvement of ecological cycles derived from the more efficient use of manure, as it is the base of key nutrients for crops. The idea of turning waste into a good instead of a problem is encouraged and financed from the point of view of research and technology. All these actions, beyond the economic and social gains to which they contribute directly, are also implemented with the aim of alleviating the environmental pressure of a sector that receives close attention because of its high pollution intensity. Estimates from the European Union calculate an emission reduction potential of between 30 and 40% (Peyraud 2016).

1.5 Conclusions

Sustainable Development Goals (SDGs), together with the Paris Agreement, are two of the major challenges facing humanity today. Livestock systems, and in particular dairy sectors, are called on to be main actors in the fulfilment of all the commitments undertaken. At a time when the world population is growing, and with 80% of the population basing their diet on the daily consumption of milk or milk products, the implementation of actions in the dairy sector that ensure a secure supply of milk, improve productivity, help the development of local communities and increasingly reduce their environmental impact is highly necessary.

Although data in recent years show how the dairy sector has greatly reduced its intensity of emissions, it has not been possible to reduce the total volume of emissions, mainly due to the increase in global herd, especially in low-productivity regions such as Africa or India. However, there are still possibilities for improvement that need to be explored. Following the guidelines proposed by FAO and GDP (2018), there are three lines of action required in the near future, taking into account the population growth scenario. Firstly, an increase in productivity must remain a key goal. All actions aimed at increasing efficiency in any section of the facilities must be encouraged. Improvements in cow diets related to nutrients and proteins, feed technology, the search for locally low-emission suppliers, the use of non-fossil fuels on farms, or animal health are among the desirable actions. Secondly, the protection of carbons sinks must be guaranteed. All livestock systems must avoid the degradation of ecosystems and deforestation. Actions related to the feed of the animals, to better grassland management and to optimised uses of low-carbon fertilizers and manure would also be good practices. Thirdly, bioeconomic solutions for livestock and dairy systems should be fostered as a holistic pathway to achieving sustainability in the sector. The implementation of a circular model in livestock systems is quite feasible and might bring a number of benefits aligned with the SDG. Recycling and reuse of energy and waste must be a foundation of the productive model of these systems.

Fulfilling the Paris Agreement commitments will require much more than reducing fossil fuel emissions in the mid-term. Solutions closely tied up with the bioeconomy will be key in the near future, because it incentivises not only the reuse and recycling of inputs, which reduces the environmental pressures, but also R&D in new fields and applications that might lead to a transformation of the technology and techniques, generating positive spillovers into society. Developed economies, such as those of the European Union, show a clear commitment to the development of the bioeconomy, not only to fulfil environmental targets but also as a potential engine of economic and employment growth. The boost to the bioeconomy in developed regions should extend its influence on developing countries through technology transfer. Such transfers are considered in the Paris Agreement to be of one of the objectives and obligations of the countries that have ratified the accord.

Livestock and dairy will continue to play an important role in the solution to the problem of production and distribution of food and nutrients globally. But now they must also be part of the solution to the climate crisis. From governments to small-size producers, there is a series of possibilities for moving towards more sustainable industry.

References

K. Anderson, G. Peters, The trouble with negative emissions. Science **354**(6309), 182–183 (2016). https://doi.org/10.1126/science.aah4567

C.L. Delgado, Rising consumption of meat and milk in developing countries has created a new food revolution. J. Nutr. **133**(11), 3907S–3910S (2003). https://doi.org/10.1093/jn/133.11.3907S

C. Devendra, Smallholder dairy production systems in developing countries: Characteristics, potential and opportunities for improvement—review. Asian-Australas J. Anim. Sci. **14**(1), 104–113 (2001). https://doi.org/10.5713/ajas.2001.104

DSF, *Dairy Sustainability Framework.* (2019). https://dairysustainabilityframework.org/2019

EC, *Directive 2001/81/EC of the European Parliament and of the Council of 23 October 2001 on National Emission Ceilings for Certain Atmospheric Pollutants*, ed. by E. Commission (2001)

EC, *Directive 2010/75/EU of the European Parliament and of the Council of 24 November 2010 on Industrial Emissions (Integrated Pollution Prevention and Control)*, ed. by E. Commission (2010)

EC, *Directive (EU) 2016/2284 of the European Parliament and of The Council of 14 December 2016 on the Reduction of National Emissions of Certain Atmospheric Pollutants, Amending Directive 2003/35/EC and Repealing Directive 2001/81/EC*, ed. by E. Commission (2016)

European Commission, *Innovating for Sustainable Growth: A Bioeconomy for Europe,* ed. by D.-G. f. R. a. I. E. Commission, (EU Publications, Brussels, 2012)

European Commission, *Bioeconomy for Europe Communication*. (European Comission Publications Office, 2015)

FAO, *The Agricultural Sectors in Nationally Determined Contributions (NDCs). Priority Areas for International Support*. Food and Agriculture Organization of the United Nations (FAO, 2016)

FAO, *Livestock Solutions for Climate Change*. Food and Agriculture Organization of the United Nations, (FAO, 2017), (p. 8)

FAO, *Policy analysis of nationally determined contributions (NDC) in Europe and Central Asia*. (Food and Agriculture Organization of the United Nations, Budapest, 2018), p. 84

FAO and GDP, *Climate Change and the Global Dairy Cattle Sector–The Role of the Dairy Sector in a Low-carbon Future*. (Food and Agriculture Organization of the United Nations and Global Dairy Platform Inc, Rome, 2018), (p. 36)

FAOSTAT, *Enteric Fermentation Dataset*, ed. by Food and Agriculture Organization of the United Nations, (FAO, 2019a)

FAOSTAT, *Livestock Primary Dataset,* ed. by Food and Agriculture Organization of the United Nations, (FAO, 2019b)

S. Frank, P. Havlík, E. Stehfest, H. van Meijl, P. Witzke, I. Pérez-Domínguez, et al., Agricultural non-CO_2 emission reduction potential in the context of the 1.5 °C target. Nature Climate Change, **9**(1), 66–72. (2019). https://doi.org/10.1038/s41558-018-0358-8

F. Fuller, J. Huang, H. Ma, S. Rozelle, Got milk? The rapid rise of China's dairy sector and its future prospects. Food Policy **31**(3), 201–215 (2006). https://doi.org/10.1016/j.foodpol.2006.03.002

GDAA, *The Dairy Sector: Ready to Help Achieve The Sustainable Development Goals*. (The Global Dairy Agenda for Action, 2015)

P. Gerber, T. Vellinga, C. Opio, H. Steinfeld, Productivity gains and greenhouse gas emissions intensity in dairy systems. Livestock Sci. **139**(1), 100–108 (2011). https://doi.org/10.1016/j.livsci.2011.03.012

GRA and CCAFS, Livestock development and climate change: The benefits of advanced greenhouse gas inventories. Global Res. Alliance Agric. Greenhouse Gas. (2016)

M. Harris, *Good to Eat: Riddles of Food and Culture*. (Waveland Press, 1998)

E.G. Hertwich, G.P. Peters, Carbon footprint of Nations: A global, trade-linked analysis. Environ. Sci. Technol. **43**(16), 6414–6420 (2009). https://doi.org/10.1021/es803496a

L. Loyon, C.H. Burton, T. Misselbrook, J. Webb, F.X. Philippe, M. Aguilar et al., Best available technology for European livestock farms: Availability, effectiveness and uptake. J. Environ. Manage. **166**, 1–11 (2016). https://doi.org/10.1016/j.jenvman.2015.09.046

H.-J. Lutzeyer, *Introduction to the European Bioeconomy Strategy*. Paper presented at the SCAR-CASA Workshop, (Carcavelos, Portugal, 2019)

J.C. Minx, T. Wiedmann, R. Wood, G.P. Peters, M. Lenzen, A. Owen et al., Input-output analysis and carbon footprint: An overview of applications. Econ. Syst. Res. **21**(3), 187–216 (2009)

U.M. Persson, D.J.A. Johansson, C. Cederberg, F. Hedenus, D. Bryngelsson, Climate metrics and the carbon footprint of livestock products: where's the beef? Environ. Res. Lett. **10**(3), 034005 (2015). https://doi.org/10.1088/1748-9326/10/3/034005

G. Peters, R. M. Andrew, J. Karstensen, *Global Environmental Footprints. A Guide to Estimating, Interpreting and Using Consumption-Based Accounts of Resource Use and Environmental Impacts*. (Copenhagen, Denmark, 2016)

G. P. Peters, R.M. Andrew, J.G. Canadell, S. Fuss, R.B. Jackson, J.I. Korsbakke, et al., Key indicators to track current progress and future ambition of the Paris Agreement. Nature Climate Change, **7**, 118. (2017). https://doi.org/10.1038/nclimate3202. https://www.nature.com/articles/nclimate3202#supplementary-information

G.P. Peters, O. Geden, Catalysing a political shift from low to negative carbon. Nature Climate Change **7**, 619 (2017). https://doi.org/10.1038/nclimate3369

J.-L. Peyraud, *The Role of Livestock in an EU-Bioeconomy*, (Animal Task Force (ATF), 2016)

Rabobank. *Dairy and the Sustainable Development Goals. The Dairy Sector's Contributions and Opportunities,* Trans. by RaboResearch, (Rabobank Industry Note #574: Rabobank, 2016)

A. Reisinger, H. Clark, How much do direct livestock emissions actually contribute to global warming? **24**(4), 1749–1761. (2018). https://doi.org/10.1111/gcb.13975

M.M. Rojas-Downing, A.P. Nejadhashemi, T. Harrigan, S.A. Woznicki, Climate change and livestock: Impacts, adaptation, and mitigation. Climate Risk Manage **16**, 145–163 (2017). https://doi.org/10.1016/j.crm.2017.02.001

O. Schmid, S. Padel, L. Levidow, The bio-economy concept and knowledge base in a public goods and farmer perspective. Bio-based Appl Econ **1**(1), 47–63 (2012)

K. Stadler, R. Wood, T. Bulavskaya, C.-J. Södersten, M. Simas, S. Schmidt et al., EXIOBASE 3: Developing a time series of detailed environmentally extended multi-regional input-output tables. J. Ind. Ecol. **22**(3), 502–515 (2018). https://doi.org/10.1111/jiec.12715

F.N. Tubiello, M. Salvatore, S. Rossi, A. Ferrara, N. Fitton, P. Smith, The FAOSTAT database of greenhouse gas emissions from agriculture. Environ. Res. Lett. **8**(1), 015009 (2013). https://doi.org/10.1088/1748-9326/8/1/015009

UN, *Kyoto Protocol to the United Nations Framework Convention on Climate Change.* (United Nations, New York, USA, 1997)

UN, The paris agreement, in *U. N. F. C. O,* ed. by C. Change, (United Nations, Paris, 2015a), (pp. 25)

UN, Sustainable development goals. (2015b). https://sustainabledevelopment.un.org/topics/sustainabledevelopmentgoals. Accessed 30 July 2017

UN, Revision of world population prospects. (2019). https://population.un.org/wpp/

A.S. Wiley, Milk for "Growth": global and local meanings of milk consumption in China, India, and the United States. Food Foodways **19**(1–2), 11–33 (2011). https://doi.org/10.1080/07409710.2011.544159

A. Wilkes, S. van Dijk, Tier 2 inventory approaches in the livestock sector: A collection of agricultural greenhouse gas inventory practices. UNIQUE forestry and land use GmbH. (2018)

P. Williamson, Emissions reduction: Scrutinize CO_2 removal methods. Nature **530**, 153–155 (2016). https://doi.org/10.1038/530153a

Chapter 2
The Ruminant: Life History and Digestive Physiology of a Symbiotic Animal

Francisco Javier Pérez-Barbería

Abstract Ruminants are the main pillar of our animal stock, and were crucial to the process of human Neolithization, as the first species to be domesticated for husbandry. They are an important element of the world's economy and cultural heritage, and also play a significant role in promoting biodiversity within the habitats they occupy. They have evolved a digestive system that relies entirely on a symbiotic relationship with micro-organisms, most of their energy comes from the end-products of microbial digestion, enabling ruminants to make use of the plant cell wall, which is something that no other vertebrate can do to such an extent. This, together with an efficient mechanism of nitrogen recycling, converts the ruminant into an efficient animal able to subsist on plant fibre, one of the most abundant organic resources in nature. Ruminants also have dental and behavioural (rumination) adaptations to comminute food and so facilitate the activity of ruminal micro-organisms, and very long intestines and caeca to increase the time food is exposed to enzymatic digestion and absorption. Brief descriptions of food energy losses and the main metabolic paths of the transformation of dietary carbohydrates, proteins and lipids are given here. Food digestion, mainly of fibre, comes at the cost of gas emissions, especially methane, which reduce food use efficiency and contribute to global greenhouse gas emissions. The purpose of this chapter is to provide a brief overview of the ruminant animal, its taxonomic diversity and life history traits, the relevance of domestication, and its adaptations to the use of plant-based diets and digestive physiology, in order to gain a better understanding of the relationships between diet and gas and solid emissions. We focus on the ruminant over monogastric species for two reasons: (i) the greater biomass contribution of ruminants to livestock, and (ii) the very complex ruminant digestive system, which includes both foregut and hindgut enteric fermentation, while monogastric species have only hindgut fermentation. Comments on dietary components and their metabolic transformations refer to roughage natural diets, rather than concentrate or supplemented diets. Although in many cases these are equivalent, we remark the importance of roughage diets because they have been the driver of the evolutionary adaptation of the ruminant symbiotic digestive system, and because of the importance of the use of roughage resources in reducing the carbon footprint of these species as compared to concentrate feeds, the production of which is high in carbon.

© The Author(s), under exclusive license to Springer Nature Switzerland AG 2020
S. García-Yuste, *Sustainable and Environmentally Friendly Dairy Farms*,
SpringerBriefs in Applied Sciences and Technology,
https://doi.org/10.1007/978-3-030-46060-0_2

2.1   Taxonomy, Evolution and Feeding Styles of the Ruminant Animal

A ruminant is any member of the mammalian order Artiodactyla (suborder Ruminantia), which is characterised by a digestive system with a multi-chambered stomach formed by rumen, reticulum, omasum and abomasum diverticula. Other anatomical traits are lack of upper incisors, presence of incisiform lower canine teeth, and a tarsus constituted by the fusion of the navicular and cuboid bones (Gentry 2000).

There are some 193 species of extant ruminants, in six families, namely Tragulidae (3 genera, 4 species), Moschidae (1 genus, 5 species), Bovidae (49 genera, 140 species), Giraffidae (2 genera, 2 species), Antilocapridae (1 genus, 1 species) and Cervidae (17 genera, 41 species) (Nowak 1999). The group is characterised by a wide inter-species range in body mass, from 2 kg in Tragulidae up to over 1000 kg in the male giraffe, and a median body mass of 45 kg (Van Wieren 1996) (Fig. 2.1).

Wild ruminants contribute to promoting biodiversity in the habitats they occupy, due to their grazing, trampling and nutrients mobilisation activities, which creates habitat heterogeneity (Gordon and Prins 2008).

Primitive ruminants have their extant representatives in the Tragulidae family (Janis 1984), which were forest dwellers with relatively simple social behaviour. Bovidae is the family with the greatest number of species, which are generally highly gregarious and with complex social behaviour, and among them were the wild species on which livestock domestication was based. It is suggested that there are five extinct families of Ruminantia: Dromomerycidae, Hypertragulidae, Gelocidae, Leptomerycidae and Palaeomerycidae (Carroll 1990), whose main habitat might have been subtropical forest with mixed clear-outs (Janis and Manning 1998). The most primitive family is Hypertragulidae, whose origins can be traced back to the

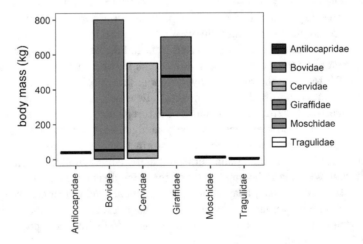

Fig. 2.1 Average, minimum and maximum body mass (kg) of the six extant families of ruminant species. Average: thick-black horizontal line; box top side: maximum, box bottom side: minimum

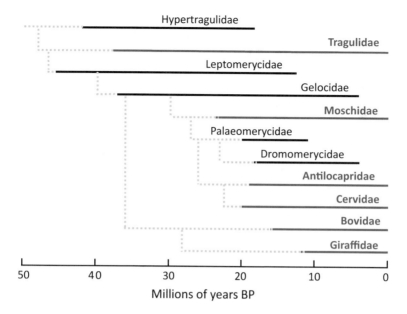

Fig. 2.2 Phylogenetic tree (units = millions of years before present) of ruminant families (red font: extant families; black font: extinct families). Solid lines represent data from fossil records; dotted lines: estimated phylogenetic relationships based on different molecular and morphological traits [adapted from Hackmann and Spain (2010)]

Early Eocene, 50 million years BP (Fig. 2.2). The extinction of these families took place between 5 and 20 million years ago.

There are no accurate estimates of the current number of wild ungulates, varying between 75 million (Hackmann and Spain 2010) and 206 million (Pérez-Barbería 2017), with Bovidae and Cervidae as the most abundant and widely distributed. Non-introduced wild ruminants occupy all continents except Australia and Antarctica, and the richest ruminant assemblages are found in Africa and Eurasia. They occupied a wide range of habitats and climates, from open to forest, with a preference for warm climates where the most diverse community of ruminants, in terms of taxonomy and morphology, has arisen (Nowak 1999; Van Wieren 1996) (Figs. 2.3 and 2.4).

Associated with their habitats are their feeding styles, a classification that characterises their dietary preferences. The traditional feeding styles are those proposed by Hofmann and Stewart (1972), who described not only dietary preferences but also the adaptations of their digestive system to diet (Clauss et al. 2010; Hofmann 1989; Pérez-Barbería et al. 2004, 2001; Pérez-Barbería and Gordon, 1999). The feeding style classifies ruminant species into (i) browsers, those species with a preference for leaves, shoots and fruits; (ii) grazers, those species with a preference for grasses; and (iii) mixed or intermediate feeders, those species that use both browsing and grazing, depending on availability and seasonal changes in plant quality. Some authors also consider the class frugivorous, which comprises the species with preference for highly nutritional and low fibre parts of plants, such as fruits (Pérez-Barbería and

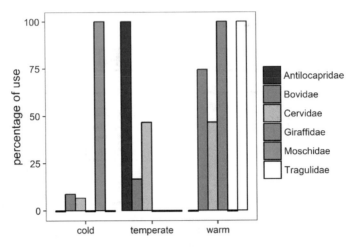

Fig. 2.3 Climates used by the six extant families of ruminants. Percentage is the number of species within a family using a particular climate. Data from Van Wieren (1996)

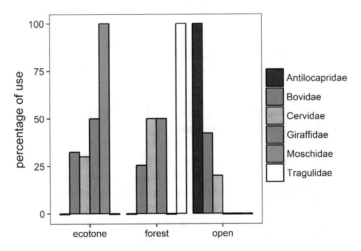

Fig. 2.4 Habitats used by the six extant families of ruminants. Data from Van Wieren (1996)

Gordon, 1999) (Fig. 2.5). Browser is the class that characterises primitive ruminants and comprises most of the evolutionary history of the Ruminantia, although most of the currently domesticated ruminants are in the grazer and mixed feeder classes. Bovidae and Cervidae are characterised by species from the three feeding styles, while the other four families are browsers (Fig. 2.5).

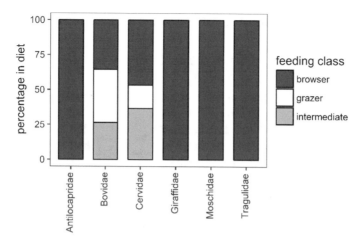

Fig. 2.5 Feeding type classification of the six extant families of ruminants. Percentage represents the proportion of the different dietary plant components. Data from Van Wieren (1996)

2.2 Domestication of the Ruminant Animal

The domestication of wild species to initiate animal husbandry was one of the keystones of Neolithization. This is the process by which primitive human populations made the transition from an economy based on hunting, fishing and gathering to an economy based on the production of plants and animals. Long term and large human settlements were limited by hunting species being locally depleted or reduced to small populations that made the provisioning of the protein necessary to sustain settlements very difficult. The domestication of ruminants provided a convenient continuous supply of milk and meat within the restricted area occupied by the settlement (Cramp et al. 2014; Gerling et al. 2017). It is claimed that the first ruminant to be domesticated was the goat, about 10,000 years ago, in the western mountains of Iran (Zeder and Hesse 2000). Husbandry of small numbers of goats made available not only milk and meat but also skins, horns and sinew for dressing and materials for tool construction. By 2,500 years BC eight ruminant species were being used for basic husbandry (goat, sheep, reindeer, water buffalo, mithan, European cattle, zebu and Bali cattle), derived from at least 12 wild ancestors that were probably gregarious species and easy to handle and breed (Hackmann and Spain 2010). Their initial use was extended to include transportation, draft and barter (Clutton-Brock 1987). Today's four main livestock ruminants add up to 4.3 billion head, and the largest population is cattle followed by sheep, goats and water buffalo (1.6, 1.4, 1.2 and 0.2 billion, respectively) (FAOSTAT 2019).

2.3   Why the Ruminant Is Different: Adaptations to Diet

2.3.1   Mouth, Teeth, Body Size and Salivary Glands

The ruminant animal is characterised by a wide range of morphological adaptations for processing and digesting the chemical compounds of the plant cell wall. Food enters the animal through the mouth, where we find the first morphological adaptations for prehension, severing and comminution of items of food. The intake and selection of food by ruminants is closely linked to the size and shape of the animal's mouth (Bell 1969; Gordon and Illius 1988; Gordon et al. 1996; Janis and Ehrhardt 1988). Browsers have narrow incisor arcades, more protruding incisors, highly mobile lips and long grasping tongues, to facilitate selection and prehension of small items of food within complex plant structures, such as leaves and shoots. Species that feed predominantly on grasses have wider muzzles and incisor arcades to maximise bite size within homogeneous grass swards; cattle are a good representative of this class. Molars have also been adapted to diets that differ in bulk, abrasiveness and mechanical resistance to comminution (Pérez-Barbería and Gordon 1998a). Grazers, in relation to body size, have a large molar occlusal surface area to grind bulky and coarse foods, and hypsodonty (high-crowned teeth) to make molars more durable and resilient to abrasion, while browsers have smaller and narrower post-canine teeth than grazers, but more prominent dental crests adapted to puncturing the cell walls of plant material (Fortelius 1985; Janis 1988; Solounias et al. 1994). Although large-bodied species are associated with some feeding styles, for example, grazers are characterised by large body size as compared with mixed feeders or many browsers, some functional adaptations remain true even after controlling for body mass (Pérez-Barbería et al. 2004). However, when the analyses take into account body mass and phylogeny, only body size and hypsodonty remained as significant adaptations to feeding styles. Browsers are smaller than mixed feeders and grazers, and browsers have shorter and less-bulky molars than those in mixed feeder and grazer species (Pérez-Barbería et al. 2004).

Another significant adaptation to the feeding style is in the salivary grands. Browse contains secondary compounds, such as phenolics, alkaloids and terpenoids, which can dramatically reduce apparent protein digestibility by binding to proteins to reduce enzymatic digestion (Robbins 1993). Browsers have well-developed mandibular and parotid salivary glands that secrete substances to neutralise the protein-binding effect of these secondary compounds, minimising the negative effect on plant fibre digestibility (Robbins et al. 1991).

2.3.2   The Ruminant Stomach

The key morphological structure that characterises ruminants is their stomach (Hofmann 1988), which is joined to the mouth by a large oesophagus. The stomach is a chamber of the ruminant digestive system that constitutes an appropriate environment for the development of a rich community of bacteria, protozoa and fungi (Karasov and Douglas 2013). Because vertebrates do not produce enzymes to digest the structural carbohydrates that constitute the plant cell wall, the ruminant relies on bacterial symbiosis to metabolise cell walls. Other highly digestible compounds such as sugars, proteins and fats can be easily metabolised by the ruminant itself via enzymatic digestion in the abomasum and small intestine. Although the stomach of any vertebrate species is colonised by micro-organisms, ruminants have evolved large-volumed multi-compartmentalised stomachs that enable a rich variety of micro-organisms to thrive in such quantities that the ruminant is not only able to use the subproducts of micro-organism metabolic activity, but its subsistence in fact entirely depends on them.

The fossil record has left no evidence of how the ruminant stomach has evolved, but its evolution can be tracked by the morphological analysis of extant primitive forms of ungulate species and herbivorous mammals. The morphological analysis focuses on the presence, size and complexity of the forestomach (Fig. 2.6).

The forestomach is the main fermentation chamber, located before the true stomach where enzyme digestion and some absorption of water and nutrients takes place. The development of the forestomach is likely to be a co-evolutionary process as a

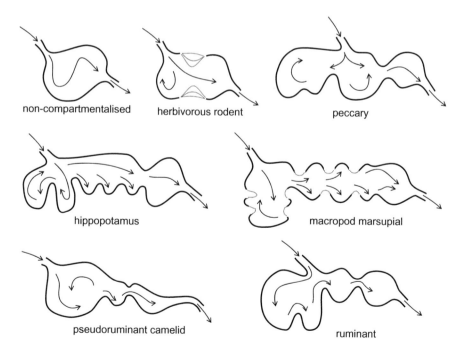

non-compartmentalised herbivorous rodent peccary

hippopotamus macropod marsupial

pseudoruminant camelid ruminant

Fig. 2.6 Different types of forestomach, from a simple one-chambered stomach to the complex four-chambered ruminant stomach. Arrows show the direction of flow of food from the oesophagus towards the intestine through the stomach [modified from Robbins (1993)]

consequence of the widespread development of grass species and grasslands in the early Miocene (Janis 1982), and of the progressive reduction of the area occupied by tropical and subtropical forests that took place from the end of the Oligocene. Grasses have a greater proportion of structural carbohydrates in their cell walls than browse (Duncan and Poppi 2008), which favours positive selection of any morphological adaptation that improved cellulose digestion in those early pre-ruminants that had adopted a diet rich in grasses, such as developing a forestomach. The reduction of tropical and subtropical forest from the end of the Oligocene contributed to changes in the feeding behaviour and patterns of aggregation of ungulates (Janis 1984; McNaughton 1984); animal aggregation promoted increasing body mass and sexual dimorphism (Pérez-Barbería et al. 2002) and probably the size of the forestomach, as it scales with a factor of 1.0 against body mass (Luna et al. 2012; Van Wieren 1996)

Simple forestomachs are those of herbivorous rodents that comprise a single diverticulum with poor differentiation from the true stomach area where enzymatic digestion occurs (Fig. 2.6). Other non-ruminant species, such as peccaries, hippopotamuses and macropod marsupials, have well-developed forestomachs (Fig. 2.6). The ruminant stomach is a structure of four chambers, namely rumen, reticulum, omasum and abomasum (Fig. 2.7).

Some authors consider rumen-reticulum to be the same functional unit. The rumen is the largest chamber, characterised by a thick wall of keratinised epithelium and muscle pillars, internally covered in papillae that increase its absorptive surface area up to 30-fold with respect to a non-papillated one (Robbins 1993). After plant material is severed, chewed and mixed with saliva, it is swallowed via the oesophagus into the rumen. Ruminants produce large quantities of saliva, rich in bicarbonate (up to 150 l/day in cattle), which buffers the huge amount of acid produced in the rumen to maintain rumen pH constant. The rumen provides an anaerobic environment with

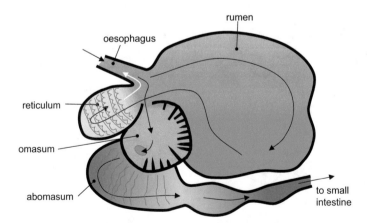

Fig. 2.7 Basic morphology of the ruminant stomach. Black arrows indicate the direction of flow of food, and the white arrow is the regurgitation of food for rumination

constant conditions that favour fermentation (pH = 6.0–6.4; temperature = 37.5–42 °C). The muscled rumen walls enable excellent peristaltic movement that allows mixing of the chewed plant material. This motility also facilitates regurgitation to the mouth of large food particles in the form of a bolus to be rechewed (rumination) to improve comminution. This cycle is repeated numerous times until food is reduced to a fine particle size, which depends on the chewing efficiency of molars, their occlusal surface area and wear (Pérez-Barbería and Gordon 1998a, 1998b). Rumination is also present in camelids and macropod marsupials, where it is called merycism. Gases, solids and liquids stratify within the rumen based on their specific gravity. On the bottom layer lies the dense and fine comminuted plant material together with liquids; on this layer the coarser less well-ground material floats like a raft, and gases rise to the top. During the fermentation process a huge amount of gas (carbon dioxide, methane and hydrogen) is produced. The more dense and fine material is conducted into the reticulum, which is a chamber with a honey-comb reticulated epithelium that forms small pools to help to retain particles of different density and size. The reticulum opens into the omasum through the reticulo-omasal orifice. The omasum is characterised by successive longitudinal walls of epithelium that further facilitate the passage of small particles of food and the attached ruminal microbes that are digesting cell walls, while retaining the larger and less digested food particles. The omasum also separates the rumen-reticulum environment from the highly acidic conditions of the abomasum where the absorption of water, soluble particles and products of the microbial activity occurs (Prins et al. 1972). The fermentation products are either absorbed through the rumen walls and delivered to the blood stream, or flow down to the small intestine for further digestion and absorption.

Camelids are considered pseudo-ruminants (dromedary and Bactrian camels, alpaca, llama, vicuña and guanaco), with a multi-compartmentalised stomach similar to that of ruminants (Fig. 2.6). Their forestomach is a truly functional fermentation chamber, as indicated by the huge amount of methane produced, similar in volume to the enteric methane yield of many ruminant species (Pérez-Barbería 2017).

2.3.3   Small Intestine

The abomasum connects with the small intestine where most of the enzymatic digestion and absorption of the nutrients that have escaped microbial action takes place. Three zones are distinguished, a proximal duodenum, an intermediate jejunum and a distal ileum. Herbivorous species, and particularly ruminants, have the longest small intestines in mammals, followed by frugivores and granivores. Species that use highly nutritious and digestible diets, such as carnivores and insectivores, have short intestines. The length of small intestine within species also depends on seasonal dietary changes (Robbins 1993; Sibly et al. 1990). The longer the small intestine the slower the passage of food through the digestive tube, which facilitates the digestion

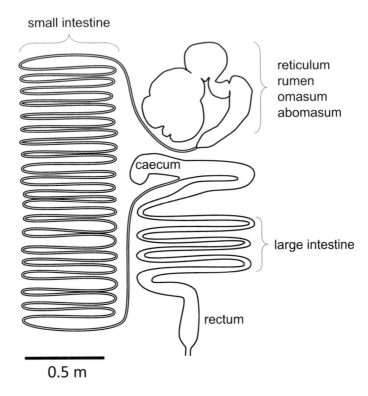

Fig. 2.8 Digestive tube of a sheep, from oesophagus to anus, depicting the relative length of each chamber and tube-like part

and absorption of plant components that are difficult to digest. Small and large intestine can add up to 12 and 30 times the body length in browser and grazer ruminants, respectively (Hofmann 1985) (Fig. 2.8).

The small intestine receives digestive secretions from four sources, liver, pancreas, intestinal mucosa and duodenal glands, which make it possible to combine a variety of enzymes, bile salts, and buffering solutions for the regulation of chemical reactions and pH. Some of these enzymes are produced as proenzymes that only become active by a series of complex reactions in the intestine. The small intestine of the neonatal ruminant has evolved adaptations that, together with enzyme inhibitors of the maternal colostrum, enable the absorption of intact molecules of colostral immunoglobulins by pinocytosis, which allows the transfer of passive immunity from the mother to the offspring (Hurley and Theil 2011). In ruminants, pseudo-ruminants and macropod marsupials, some saccharidases are missing, as many disaccharides are fermented in the forestomach to volatile fatty acids. Similarly, fat digestion by pancreatic lipase and bile salts in the forestomach is limited, as microbes hydrolyse fatty acids of food glycerides and ferment the glycerol fraction (Robbins 1993).

2.3.4   Caecum and Large Intestine

Caeca are diverticula located at the junction between the small and large intestine. In ruminants there is only one caecum, which is a large and functional digestive structure. The caecum opens to the large intestine at the ileo-caecal juncture and ends at the rectum. As with the small intestine, a long large intestine reduces the rate of food passage, improving digestion and absorption of plant material resilient to digestion. The large intestine is up to 78% of the small intestine length (Barry 1977) (Fig. 2.8). Further fermentation of fibre, endogenous matter, and absorption of water and water-soluble nutrients, amino acids, volatile fatty acids, ammonia, electrolytes and bacterial synthesis of vitamins all take place in the caecum and large intestine. At the same time, the caecum and large intestine select remaining matter from the digestion and prepare it for the formation of faeces (Robbins 1993).

2.4   Cattle: A Large Contributor to Greenhouse Gas (GHG) Emissions

Special importance is given in this book to cattle because of their significant contribution to GHG, due in turn to their large body size and a world population that in 2017 exceeded 1.5 billion million head (FAOSTAT 2019). A brief overview of cattle taxonomy and natural habitat and diet is provided here.

2.4.1   Cattle Taxonomy, Habitat and Diet

Cattle were the first domesticated large ruminants, about 8,500 years BC (Scheu et al. 2015) in the same geographical area where goats, sheep and pigs are also claimed to have been domesticated (Larson and Burger 2013). Cattle are our main source of milk, and together with other ruminants (sheep and goats) and pigs are an important source of meat and leather across the world, and in many countries, cattle are still a useful tool to provide power and traction.

Cattle (*Bos taurus*) is an artiodactyl of the Bovidae family, which comprises antelopes, cattle, bison, buffalo, goats and sheep, with 49 extant genera and 140 species, distributed across Africa, most of Eurasia, North America, and some islands of the Artic and East Indies. The genus *Bos* has five species: *B. taurus*, which includes the extinct aurochs and domestic cattle, originally distributed across Europe, Asia (except the far north and southeast) and northern Africa; and *B. sauveli*, *B. javanicus*, *B. gaurus* and *B. grunniens* (Nowak 1999). It is accepted that in Europe two wild cattle species have co-existed, namely the European bison (*Bison bonasus*) and aurochs (*Bos primigenius*). The ancestor of domesticated cattle was the aurochs, which was first used as a game species and was then domesticated. The aurochs was likely to spread into Europe through India (Thenius 1980); they were larger than modern cattle, with a plausible shoulder height of 160–180 cm in bulls and 150 cm in the cow, estimated using scaling regressions of current and ancient bones. Domestication of the aurochs started in the Middle East and Pakistan about 9,000 years ago (Badam 1984; Troy et al. 2001), and this process has produced about 1,000 cattle breeds (Felius 2007; Porter et al. 2016), in which body size and sexual dimorphism reduction have been the most important morphological changes (Grigson 1978, Bohlken 1964 in Van Vuure 2005), other than improvement of meat and milk yield and draught activity. Some old cattle breeds, such as Spanish fighting bulls, retain some morphological traits of the aurochs (Van Vuure 2005).

The aurochs became extinct in southern Asia and northern Africa 2,000 years ago, but survived in western and central Europe until 1400. The last known aurochs lived in Poland in the Jaktorów forest and became extinct in 1627 (Van Vuure 2005). Nowadays, it is accepted that *Bos primigenius* and *Bos taurus* are the same species, although some authors consider they should be taxonomically differentiated.

Reports from the time that the aurochs became extinct (Gesner 1602 in Van Vuure 2005) indicate they were social animals that gathered in large groups in winter, while in summer there were groups of cows with calves and young bulls, segregated from older bulls and solitary bulls. The aurochs mating season was from August to September, and calves were born in May or June.

Based on palynology studies it is believed that the aurochs lived during the landscape transition that took place in Europe around 3000 BC, in which open habitats started to dominate woodland (Van Vuure 2005). Its original habitat probably resembled that of other extant forest-dwelling bovids like the forest buffalo (*Syncerus caffer*) and the wood buffalo (*Bison bison*), together with grazing on sedge marshes and riverbanks. Descriptions by Gesner (1602, in Van Vuure 2005) suggest that the aurochs diet was that of a typical grazer, similar to modern cattle, feeding mainly on grasses and graminoids, including acorns, and browsing from bushes and trees in autumn and winter.

Cattle lie at the far end of the grazer high-fibre feeding style, relying on the efficient digestion of cellulose, for which they have a large rumen-reticulum capacity to accommodate large amounts of fibrous plant material to make it possible to increase food retention for efficient fermentation.

2.5 Ruminant: The Efficient Animal

There is a large number of domesticated terrestrial species that contribute to GHG emissions, among which are asses, bees, buffaloes, camelids, camels, cattle, chickens, ducks, geese, guinea fowl, goats, horses, mules, pigeons, pigs, rabbits, hares, rodents, turkeys and sheep (FAOSTAT 2019). But the main source of GHG, particularly in methane emission, is produced by enteric fermentation in the ruminant animal (Janssens-Maenhout et al. 2017; Wuebbles and Hayhoe 2002), and not only from the domesticated group of species but also from the 209 million animals that constitute the wild populations of ruminant species (Nowak 1999; Pérez-Barbería 2017).

The ruminant animal is characterised by being able to subsist entirely on plant components, a unique adaptation for vertebrates of body mass over 1000 kg. Other non-ruminant large herbivores are the hindgut fermenters (e.g. pigs, equids, proboscideans, hippopotamuses), which can also subsist on plant material, but their digestive efficiency is lower than that in ruminants (Church 1988).

The work produced per unit of gross energy of food ingested measures the food use efficiency of an animal, and the higher the efficiency, the lower the emissions of solids and gases to the environment. There are other factors that have to be considered to produce a fair comparative inter-species estimate of food efficiency use and emissions from digestive processes. Foods with similar gross energy value might differ greatly in the amount of energy required to produce them; as a general rule, the higher the level in the trophic chain the food comes from, the longer and less efficient is the energy path that produces it. The ruminant is an efficient animal because it is able to use the energy stored in the tissues of the primary producers of the trophic chain, and because of its symbiotic relationship with microbes, it is able to utilise not only plant cell components but also cell walls (i.e. plant fibre) that are very resilient to digestion by most non-ruminant mammalian species. Plant fibre is one of the most abundant types of organic matter on Earth, which is produced in a sustainable way without any human intervention, and ruminants can subsist feeding on fibrous diets. This makes ruminants efficient animals for producing meat, milk, leather and wool for human consumption.

Despite the ruminant's high digestive efficiency, there is a considerable amount of solid and gas by-product from the digestion, which are used in estimating the carbon footprint of the animal. Although an over-simplification of the complex flow of energy through the ruminant, a widely accepted partitioning of the energy is as shown in Fig. 2.9 (AFRC 1993; Ferrel 1988; National Research Council 2006).

Energy input occurs via food intake, and the amount of energy provided is called gross energy. The amount of energy produced by the food ingested can be measured in a bomb calorimeter by combustion, as the amount of heat produced by complete oxidation of the chemical compounds. However, the energy an animal can extract from most feeds is not at all close to gross energy. Gross energy depends on the proportion of carbohydrate, protein and fat in diet, as their average heat combustion is from 3.7 (glucose) to 4.2 (starch), 5.6 and 9.4 kcal/g, respectively. The first loss of energy occurs as faecal energy, that in ruminants can be as high as 40–50% of dry matter intake when fed forage, or 20–30% when fed a high-concentrate diet. The remaining energy is apparent digestible energy. It is called apparent because faecal energy not only contains undigested food but also non-dietary components, such as animal internal metabolic products, ruminal microbes and epithelial cells of the rumen, small and large intestine. Another source of energy loss (2–5%) is via urine in urea and other non-urea nitrogenous compounds, such as hippuric acid, which are highly dependent on the type of diet. Diets rich in plant secondary compounds increase urinary energy losses (Robbins 1993). Combustible gases, mainly methane, produced during ruminal fermentation are belched, and a small proportion from

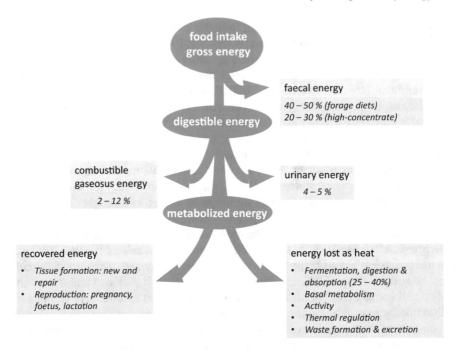

Fig. 2.9 Energy partitioning diagram from gross energy provided by food intake to recovered energy, together with some percentages of energy losses (National Research Council, 2006)

hindgut fermentation is expelled through the anus. These gases can account for between 5 and 8% of digestible energy in ruminants, and in cattle it can be as high as 12% (Johnson and Johnson 1995; National Research Council 2006; Pérez-Barbería 2017). Production of combustible gases from enteric fermentation increases with the proportion of forage in diet, and decreases in high-concentrate diets, and diets rich in browse produce fewer gases compared with grass diets (Robbins 1993; Pérez-Barbería et al. 2020). The remaining energy after urine and combustible gas losses is the metabolised energy, which is a good proxy for the recovered energy and energy lost as heat. Recovered energy is that used by the animal in new tissue formation, such as growth, or in processes of tissue repair, and in reproduction (pregnancy, foetus and lactation). Heat losses are those from basal metabolism, activity (e.g. respiration, standing, walking), digestion and absorption, fermentation, thermal regulation, waste creation and excretion.

In the ruminant animal the amount of retained energy increases as feed intake increases, but not in a linear fashion, as the rate is constrained by a number of factors (Ferrel 1988). In general and for a particular type of food, energy use decreases as the rate of passage of food through the stomach and intestine increases, so any factor that produces an increment in the rate of passage can have a detrimental effect on energy utilisation. For example, metabolised energy per unit of food intake is higher at low intakes as compared to *ad libitum*, and passage rate increases with intake (Meyer et al. 2010; National Research Council 2006). The chemical composition of faecal losses varies widely depending on diet, with faecal structural components between 2%, in grain processed diets, and 90% in straw-rich diets of very low digestibility. Differences in digestibility vary more between species and feeding styles than intra-species or between breeds (Clauss et al. 2013; Edouard et al. 2008; Pérez-Barbería et al. 2004).

2.6 Use of Structural and Non-structural Plant Cell Components

As mentioned above, gas production from enteric fermentation and excreted metabolic and undigested dietary components depends strongly on the type of diet. One of the main differences between the ruminant and other non-ruminant herbivorous mammals is the use of the structural components of the plant cells. Ruminants ingest mixtures of plants that are moistened in saliva, chewed, ingested, ruminated and exposed to enzymatic digestion by both symbiotic micro-organisms and the ruminant. Once plant intra-cell components are exposed by the mechanical action of mastication (Pérez-Barbería and Gordon 1998a) energy becomes available in two main components that can be classified as to how easily their energy can be used. Rapid-release energy comes from short carbohydrate polymers, proteins and fat dissolved in the cell cytoplasm, and cell membrane and organelles. On the other hand, there is a substantial amount of energy, more resilient to enzymatic digestion in vertebrates, contained in the chemical bonds of the structural carbohydrates that constitute the cell wall. These are long-chain polysaccharides, such as cellulose, composed of hundreds to thousands of units, matrix polysaccharides (hemicellulose), heteropolysaccharides and cross-linked phenolic polymers (pectin, lignin, suberin, cutin). The breaking down of these polymers requires the action of symbiotic ruminal micro-organisms. The ruminal bioma is a highly competitive environment, in which different populations of micro-organisms thrive or decline depending on the highly variable conditions that occur during digestion (Karasov and Douglas 2013). Intake normally takes place in bouts, and the digestion of one batch of food overlaps with the ingestion of a new batch, which might also vary in its dietary components. In addition, water intake and saliva are also provided at intervals, which adds more variability to the system. Despite these changes in the micro-environment, the animal provides favourable physical conditions for micro-organism populations to grow, maintaining moisture, homeothermy, stable pH range and continuous input of micro-organisms that come attached to the food and water ingested.

The animal competes with the micro-organisms for the easily accessible energy provided by the non-structural components of the plant cells, so part of this energy can be used by ruminal microbes before it can be absorbed in the abomasum or intestine. Microbes occupy three different rumen environments, depending on their respiration type (Owens and Goetsch 1988). Some live attached to the rumen wall and are tolerant of small amounts of O_2 that enter the rumen from capillary blood. These microbes can hydrolyse the urea that diffuses into the rumen and oxidise some chemical compounds. Other micro-organisms adhere to food particles that float in the rumen or are attached to protozoa that swim in the rumen, and the third group move freely or are suspended in the rumen fluid. Although the rumen is an anaerobic environment, in addition to the limited O_2 capillary blood, small quantities of O_2 are attached to food particles or diffused in the ruminal fluid. As a consequence, there are very small populations of facultative anaerobes that can use the limited O_2 available, the largest populations are from obligate anaerobic micro-organisms.

In this ruminal anaerobic environment, there is a surplus of reducing equivalents, such as NADH (a reduced form of nicotinamide adenine dinucleotide, which is an important anaerobic metabolic cofactor). This and other types of reducing equivalents are used in a variety of types of redox reactions to reduce all available compounds immersed in the ruminal environment. For example, CO_2 is reduced to methane; sulphates and nitrates are reduced to sulphides and ammonia; and unsaturated fatty acids are reduced to their saturated forms. The main products of the enteric fermentation are volatile fatty acids VFA, methane, CO_2, ammonia–nitrogen, which are continuously removed from the rumen and used by the animal as a source of energy (VFA) or disposed of as waste products of digestion, such as methane (Table 2.1). Microbe growth is limited by ATP availability, which could be increased by the use of O_2, but then the end-products of digestion would be CO_2 and H_2O without producing VFA, which are the main sources of energy for the ruminant. This highlights the fine equilibrium existing among the micro-organisms population and between the micro-organisms and the ruminant.

Table 2.1 Enteric metabolic transformation of main dietary components into final products of digestion Modified from Owens and Goetsch (1988)

Dietary component	Polymer chemical component	Monomer chemical component	Digestion products
Nitrogen-free extract	Long-chain carbohydrates (hexosan)	Glucose and other hexoses	Volatile fatty acids (acetate, propionate, butyrate)
Crude fibre	Long-chain carbohydrates (pentosan)	Pentoses	Acetate, propionate, butyrate
Crude protein	Protein, non-protein N	Amino acids	Acetate, propionate, butyrate, isobutyrate, isovalerate, ammonia
Crude fat	Triglycerides, galactosides	Fatty acids, glycerol	Propionate, saturated fatty acids
Ash	Minerals	Elements	Reduced elements, microbial cells, CO_2, methane

2.7 Carbohydrate Metabolism

Carbohydrates constitute the main source of energy in the ruminant diet. Carbohydrates can be dietary components in the form of monosaccharides, oligosaccharides and polysaccharides (Table 2.1). They have three main functions: (i) providing energy to the ruminal microbes, (ii) providing energy to the host animal, and (iii) contributing to the creation of the ruminal physical environment where the microbes can develop; this is especially important in the case of structural carbohydrate polymers. The main source of energy for the animal and end-product of fermentation is VFA (Fig. 2.10).

Non-structural polymer carbohydrates, such as starch that plants use as a store of energy, can be rapidly used by microbes or escape from them and be enzymatically hydrolysed and absorbed in the small intestine (National Research Council 2006). Cellulose is the main plant structural carbohydrate of the ruminant diet; 1 mol of glucose from cellulose, if completely oxidised, produces 38 mol of adenosine triphosphate (ATP, a chemical compound that provides energy to drive many metabolic processes). But if 1 mol of glucose is reduced by anaerobic bacteria, it produces only between 2 and 6 mol of ATP, making available to the ruminant 32–36 mol of ATP, that the animal can use via oxidative metabolism of volatile fatty acids (Robbins 1993). This explains the huge amount of energy available to the ruminant from its symbiotic relationship with the gut microbes.

More than 70% of ruminant energy is provided by VFA, and between 50 and 70% of the digestible energy ends up as VFA (National Research Council 2006). Volite fatty acids are short-chain fatty acids with less than six carbon atoms. The three more

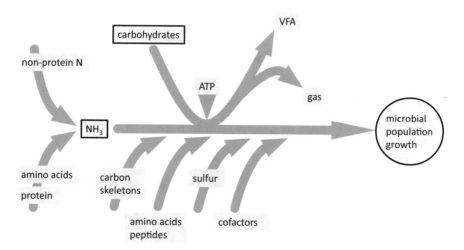

Fig. 2.10 Flow diagram of nitrogen and energy use for ruminal microbial population growth [modified from Owens & Zinn (1988)]

common VFA are acetate, propionate and butyrate (2, 3 and 4 carbon atoms, respectively), which are continuously removed through the rumen epithelium, by gradient concentration, and carried by ruminal veins to the portal vein reaching the liver. Its continuous removal from the rumen prevents low levels of ruminal pH, which would compromise fermentation rates. Although the ruminal epithelium is not designed for efficient absorption, its structure allows rapid absorption rates for VFA, water, electrolytes and lactic acid. There is also feedback between concentration of VFA in rumen and absorption capacity through the ruminal wall, as high concentrations of VFA in the rumen promote the development of size, length and density of ruminal papillae. Different VFAs undergo different metabolic transformations. Acetate and propionate are not modified when they pass through the rumen wall, but butyric acid is metabolised to beta-hydroxybutyric acid. Acetic acid does not undergo liver transformations, but is oxidised in the tissues to generate ATP, and as the principal source of acetyl CoA (i.e. a key molecule in protein, carbohydrate and lipid metabolism for energy production) in the synthesis of lipids. Propionic acid is captured by the liver from the portal blood and used as a major substrate for gluconeogenesis, and this is of paramount relevance in the ruminant, as most glucose cannot bypass rumen fermentation and reach the small intestine for absorption. Butyric acid, after being transformed into beta-hydroxybutyric acid, is oxidised in the tissues as a source of energy (Fig. 2.11).

Diets can vary widely across species of herbivorous mammals (Pérez-Barbería et al. 2004). Some small ungulates feed on fruits and shoots with high content in monosaccharides, while large grazers feed on diets rich in structural carbohydrates. In intensive production systems ruminants are usually fed on grain-based diets rich

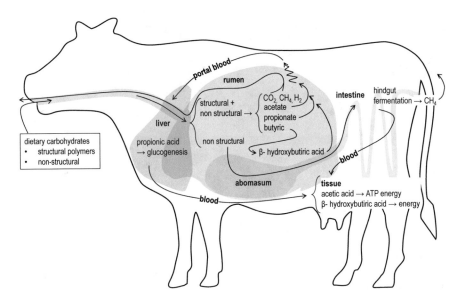

Fig. 2.11 Flow diagram of carbohydrates use by the ruminant

in starch. For proper ruminal fermentation there has to be a fine-tuning between the proportion of soluble and structural carbohydrates in the diet. Slowly digestible structural carbohydrates produce acetate as the main end-product of fermentation, while easily digestible soluble carbohydrates favour the production of propionate. Propionate supply can increase fermentation rate and the incorporation of rumen ammonia into microbial protein but at a cost of decreasing ruminal pH with the associated detriment in the fermentation rate of structural carbohydrates (Dove and Milne 1994; National Research Council 2006).

2.8 Protein Metabolism

Ruminants make use of two sources of protein, dietary protein and microbial protein. Proteins from plant-based diets are those associated with the cell and cell-organelle membrane, and those that are free in the cell cytoplasm. The ruminant gut harvests microbes from ruminal populations to scavenge its microbial protein constituents, among other chemical compounds. Because between 3 and 10% of the dry matter in ruminal microbes is protein, ruminants can subsist without a source of dietary protein (National Research Council 2006). The ruminant competes with its symbiotic microbes for dietary protein.

The levels of protein degradation in the rumen are highly variable and depend on the composition of the microbial population assemblage and the levels of nitrogen available in the rumen. Between 20 and 100% of dietary protein can be converted into ammonia in the rumen, while the remaining fraction will escape from microbial digestion and will reach the small intestine for digestion. The hydrolysis of proteins in the rumen is a complex process as the end-products depend on the type of microbes involved in their degradation (Fig. 2.12).

Insoluble proteins are first solubilised in contact with molecules that expose their hydrophilic amino acids. Proteolytic enzymes (endo- and exo-proteases and pectidades) cleave peptide bonds, releasing a mixture of amino acids and peptides of varying size. This hydrolysis occurs on the surface of the microbial cell wall, where these proteolytic molecules are attached. The amino acids and peptides released are rapidly used by microbes, either in their original form or removing their amino group with deaminase enzymes to produce energy and ammonia. The use of peptides and amino acids by microbes is so fast that their levels in rumen are very low. Those proteolytic bacteria that cannot make use of amino acids use ammonia as a source of nitrogen, while peptides are used as a source of carbon and energy (Fig. 2.10).

The dietary protein that escapes from the action of microbes passes into the small intestine where it is digested and absorbed. Similarly, microbial crude protein passes through the omasum and abomasum to the small intestine for enzymatic digestion by pancreatic enzymes, pepsin and HCl. Both crude microbial protein and dietary protein have rates of 65–75% of post-ruminal N digestion in the duodenum (Owens and Zinn 1988). Protein that is resilient to microbe digestion and passes into the small intestine is likely to have low rates of duodenal digestion too, but in general,

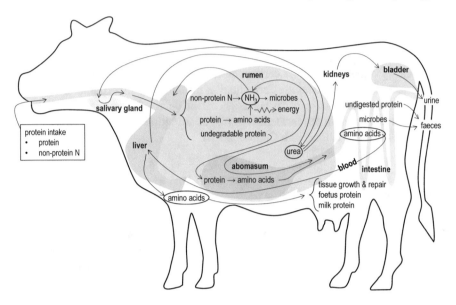

Fig. 2.12 Flow diagram of protein and nitrogen use by the ruminant

the digestive and absorptive protein capacity of the jejunum and ileum is very high, especially in the absorption of L-form amino acids that can be used by cells directly.

Amino acids cannot be stored in tissue, therefore the amino acids that are released from protein in the small intestine have different fates: (i) they are used as the main building blocks in protein synthesis, because ruminants cannot synthesise the carbon chain of the essential amino acids, (ii) their carbon skeletons can be a source of glucose, mainly when they come from alanine, aspartate and glutamate, or (iii) they may be catabolised, where the amino-N group is converted into urea, and the carbon skeleton oxidised to CO_2. Unused protein N is excreted in the form of NH_3-N and urea in the urine. Among other losses of N in faeces are the N associated to fibre residues, microbial particles (mainly from cell walls and a smaller fraction from cell contents) and N from protein metabolism in the animal tissues. Excess of non-protein N cannot be processed, and is therefore useless and excreted. The increase in the quantity of fermentable food particles in the large intestine and caecum also increases faecal N but decreases urinary N (Fig. 2.10).

Ruminants are unique as they can digest up to 80% of microbial nucleic acids using pancreatic ribonucleases. This digestion contributes to conserving P and N; the pyrimidines produced can be absorbed into tissue and catabolised in the liver, and purines excreted in the urine.

Microbial synthesis of protein from non-protein dietary nitrogen and dietary protein is restricted by limitation of nitrogen and energy supply to the microbes from fermentation (in the form of ATP or digestible organic matter) and the efficiency of microbes at using different sources of energy (Fig. 2.10). Microbial preference for the source of nitrogen depends on species; some prefer ammonia, while others use amino acids or peptides, and protozoa use nitrogen derived from bacteria, fungal or dietary protein (National Research Council 2006).

Ruminants have evolved as an efficient mechanism of nitrogen recycling that makes it possible for them to survive on diets poor in nitrogen (Robbins 1993). Between 40 and 80% of urea-N synthesised in the liver is returned to the rumen from the blood stream by diffusion through the ruminal wall, for reuse in the synthesis of microbial protein (Lapierre and Lobley 2001). A second route for urea recycling is via saliva, which can account for between 15 and 50% of the total urea recycled (Owens and Zinn 1988). The highest levels of urea recycling correspond to roughage diets, as saliva production increases in diets rich in structural carbohydrates, and at low concentrations of ammonia in the rumen.

2.9   Lipid Metabolism

Herbivorous mammals ingest a diverse group of lipids, and generally in small quantities because of the nature of their roughage diets. Most of the non-dietary lipids are VFA of microbial production in the rumen.

Lipids in plants have two functions, structural and energy storage, and depending on their function, can be located in different parts of the plant cells (Table 2.2) (Byers and Schelling 1988). Structural lipids are constituents of the membranes of cell and cell organelles, such as chloroplasts, and also as the main compounds of plant cuticle, which is a protecting layer that covers the epidermis of leaves, stems, flowers and other parts of the plant. They can account for between 3 and 10% of the plant's dry matter. Among the main structural lipids in cell membranes are phospholipids, glycolipids, chlorophylls, carotenoids, sterols and acylated sterol glycosides. Glycolipids constitute between 40 and 50%, and chlorophylls 20%, of the membrane lipids. The lipids of plant cuticles are a variety of waxes, among them are long-chain carbohydrates (n-alkanes), very long-chain fatty acids, long-chain alcohols and aldehydes (fatty alcohols and fatty aldehydes), ketones, esters and polyester polymers such as cutin.

Plants store energy in the form of carbohydrates, as in many grains, and lipids, as in seeds like sunflower. Lipids provide about twice the amount of energy as do carbohydrates and protein (National Research Council 2006), but some waxy components of plants are very resilient to ruminal digestion, such as the long-chain n-alkanes (Dove and Mayes 1996). The lipids with energy storage function are triglycerides of chain length of between 14 and 18 carbon atoms (Table 2.2), and among the most important for the herbivores are myristic, palmitic, palmitoleic, stearic, oleic, linoleic and linolenic.

Dietary lipids are used quickly by ruminal microbes, so little fat bypasses the rumen for enzymatic digestion in the small intestine. Dietary lipids in the form of esterified fatty acids are first hydrolysed by ruminal action to produce free unsaturated fatty acids and glycerol, but protozoa and some bacteria have no capacity for lipid hydrolysis. The second step is the hydrogenation of the unsaturated fatty acids, with

Table 2.2 Function and plant location of the main lipids found in natural diets of herbivorous mammals

Function	Location	Chemical compound
Structural	Cell membranes	Phospholipids Glycolipids Chlorophylls Carotenoids Sterols Acylated sterol glycosides
	Epidermis waxy cuticle	n-alkanes (long-chain carbohydrates) Very long-chain fatty acids Long-chain alcohols (fatty alcohols) Long-chain aldehydes (fatty aldehydes) Ketones Esters Polyester polymers (cutin)
Energy storage	Cell	Triglycerides (fatty acids) • Myristic • Palmitic • Palmitoleic • Stearic • Oleic • Linoleic • Linolenic

the incorporation of H into double bonds, and this also helps to dispose of the large amount of H from the reducing rumen environment. Hydrogenation is carried out in the rumen by bacteria and also by protozoa that are very active in biohydrogenation (Byers and Schelling 1988). The required linoleic and alpha-linolenic essential fatty acids seem to bypass rumen without chemical transformations. Saturated and non-completely saturated fatty acids pass to the intestine as particulate matter, then are emulsified and integrated into micelles by the action of pancreatic lipase enzymes and bile salts. In this form it can diffuse into the intestinal cells and be reesterified to be moved into the lymphatic system by the aid of very low and ultra-low-density lipoproteins. Short-chain VFAs reach the liver directly from the rumen via portal blood. There is little fat synthesis in the liver, as most takes place in adipose tissue. The main precursors of fat reserve tissue are butyrate and lactate, and acetate is mainly used in the mammary glands. Fat reserve tissue is of paramount importance for the survival of wild and domestic ruminants in extensive systems during periods of food scarcity (Fig. 2.13).

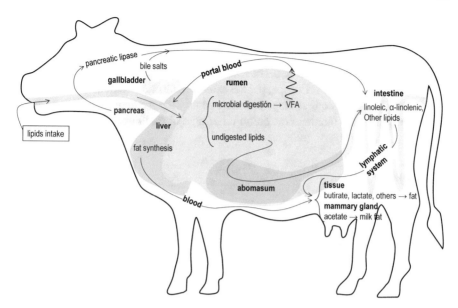

Fig. 2.13   Flow diagram of lipid use by the ruminant. VFA, volatile fatty acids

2.10   Conclusion

The ruminant is an efficient animal, as it is able to use plant fibre as its main source of energy. This is possible because of its symbiotic relationship with ruminal micro-organisms, from which digestion end-products, mainly volatile fatty acids, and microbial populations, are harvested as means of energy. Other sources of energy come from the dietary components that escape from microbial digestion, such as non-structural carbohydrates, proteins and lipids from cell membranes. Though in nature plant fibre is widely available to the ruminant, dietary protein is generally limited. Ruminants have evolved as an efficient mechanism of nitrogen use through ammonia from micro-organism activity, recycling urea and the ability to digest microbial nucleic acids using pancreatic ribonucleases.

Despite the digestive efficiency of the ruminant, gas and solid emissions are inevitably produced as a consequence of digestion, including methane, an important greenhouse gas with 25 times the warming potential of CO_2. Fibrous diets increase methane emissions, and there have been attempts to reduce methane emissions by dietary modifications and supplements, but at the cost of using dietary components whose production has a high carbon footprint, and poor animal welfare outcomes. Fibrous natural diets have been the key for the evolution of the digestive system of these species, and so they should be taken into account when attempting to balance animal welfare, production and emissions.

In recent years, the crucial role of ruminants in sustainable extensive farming in the biodiversity of these systems, maintaining grassland habitats by reducing scrub encroachment and increasing habitat heterogeneity has been acknowledged. Furthermore, there is evidence that diets rich in certain plants found on some extensive animal farming systems (heathlands) reduce methane emissions per unit of digestible organic matter. This, together with the recent interest in slowing the abandonment of deprived rural areas across Europe, put the extensive farming systems at the forefront of potential low emissions strategies, which could help the economy of rural areas, and at the same time maintain the cultural heritage of traditional animal farming.

References

AFRC, *Energy and protein requirements of ruminants. Agricultural and Food Research Council* (CAB International, Wallingford, Oxon, UK, 1993)

G.L. Badam, Holocene faunal material from India with special reference to domesticated animals, in *Animals and Archaeology 3. Early Herders and Their Flocks*. (BAR International, 1984), pp. 339–353

R.E. Barry, Length and absorptive surface area apportionment of segments of the hindgut for eight species of small mammals. J. Mammal. **58**, 419–420 (1977)

R.H.V. Bell, *The Use of the Herbaceous Layer by Grazing Ungulates in the Serengeti National Park, Tanzania*. (University of Manchester, 1969)

F.M. Byers, G.T. Schelling, Lipids in ruminant nutrition, in *The Ruminant Animal. Digestive Physiology and Nutrition*, ed. by D.C. Church (Department of Animal Science, Oregon State University, Corvallis, OR, USA, 1988), pp. 298–312

R.L. Carroll, *Vertebrate Paleontology and Evolution*, 1st Edn. ed. W. H. Freeman. (Company, New York, NY, 1990)

D.C. Church, *The Ruminant Animal. Digestive Physiology and Nutrition* (Prentice Hall, Upper Saddle River, NJ, USA, 1988)

M. Clauss, I. Hume, J. Hummel, Evolutionary adaptations of ruminants and their potential relevance for modern production systems. Animal **4**, 979–992. (2010). https://doi.org/10.1017/S1751731110000388

M. Clauss, P. Steuer, D.W.H. Müller, D. Codron, J. Hummel, Herbivory and body size: allometries of diet quality and gastrointestinal physiology, and implications for herbivore ecology and dinosaur gigantism. PLoS One **8**, e68714 (2013)

J. Clutton-Brock, *A Natural History of Domesticated Mammals*. (Cambridge University Press , British Museum (Natural History), Cambridge , London , 1987)

J.E. Cramp, R.P. Evershed, M. Lavento, P. Halinen, K. Mannermaa, M. Oinonen, J. Kettunen, M. Perola, P. Onkamo, V. Heyd, Neolithic dairy farming at the extreme of agriculture in northern Europe. Proc. R. Soc. B Biol. Sci. **281**, 20140819 (2014). https://doi.org/10.1098/rspb.2014.0819

H. Dove, R.W. Mayes, Plant wax components: a new approach to estimating intake and diet composition in herbivores. J. Nutr. **126**, 13–26 (1996)

H. Dove, J.A. Milne, Digesta flow and rumen microbial protein production in ewes grazing perennial ryegrass. Aust. J. Agric. Res. **45**, 1229–1245 (1994). https://doi.org/10.1071/ar9941229

A.J. Duncan, D.P. Poppi, nutritional ecology of grazing and browsing ruminants, in *The Ecology of Browsing and Grazing, Ecological Studies*. (Springer, Berlin, Heidelberg, 2008), pp. 89–116. https://doi.org/10.1007/978-3-540-72422-3_4

N. Edouard, G. Fleurance, W. Martin-Rosset, P. Duncan, J.P. Dulphy, S. Grange, R. Baumont, H. Dubroeucq, F.J. Pérez-Barbería, I.J. Gordon, Voluntary intake and digestibility in horses: effect of forage quality with emphasis on individual variability. Animal **2**, 1526–1533, (2008)

FAOSTAT, *Food and Agriculture Organization of the United Nations. FAOSTAT Database*. (FAO, Roma, Italy, 2019)

M. Felius, *Cattle Breeds: An Encyclopedia*. (Trafalgar Square Publishing, 2007)

C.L. Ferrel, Energy metabolism, in ed. by D.C. Church, *The Ruminant Animal. Digestive Physiology and Nutrition*. (Department of Animal Science, Oregon State University, Corvallis, OR, USA, 1988), pp. 250–268

M. Fortelius, Ungulate cheek teeth: developmental, functional and evolutionary interrelations. Acta Zool. Fenn. **180**, 1–76 (1985)

A.W. Gentry, The ruminant radiation, *Antelopes, Deer, and Relatives: Fossil Record, Behavioral Ecology, Systematics, and Conservation* (Yale University Press, New Haven, CT, 2000), pp. 11–25

C. Gerling, T. Doppler, V. Heyd, C. Knipper, T. Kuhn, M.F. Lehmann, A.W.G. Pike, J. Schibler, High-resolution isotopic evidence of specialised cattle herding in the European neolithic. PLoS ONE **12**, e0180164 (2017). https://doi.org/10.1371/journal.pone.0180164

I. Gordon, A. Illius, Incisor arcade structure and diet selection in ruminants. Funct. Ecol. **2**, 15–22 (1988). https://doi.org/10.2307/2389455

I.J. Gordon, A.W. Illius, J.D. Milne, Sources of variation in the foraging efficiency of grazing ruminants. Funct. Ecol. **10**, 219–226 (1996). https://doi.org/10.2307/2389846

I.J. Gordon, H.H.T. Prins (eds.), *The Ecology of Browsing and Grazing, Ecological Studies* (Springer-Verlag, Berlin Heidelberg, 2008)

C. Grigson, The craniology and relationships of four species of Bos,: 4. The Relationship between Bos primigenius Boj. and *B. taurus* L. and its implications for the Phylogeny of the Domestic Breeds. J. Archaeol. Sci. **5**, 123–152 (1978). https://doi.org/10.1016/0305-4403(78)90028-6

T.J. Hackmann, J.N. Spain, Invited review: Ruminant ecology and evolution: Perspectives useful to ruminant livestock research and production. J. Dairy Sci. **93**, 1320–1334 (2010). https://doi.org/10.3168/jds.2009-2071

R.R. Hofmann, Evolutionary steps of ecophysiological adaptation and diversification of ruminants—a comparative view of their digestive-system. Oecologia **78**, 443–457 (1989)

R.R. Hofmann, Anatomy of the gastro-intestinal tract, in ed. by D.C. Church, *The Ruminant Animal. Digestive Physiology and Nutrition*. (Prentice Hall, Englewood Cliff, New Yersey, 1988), pp. 14–43

R.R. Hofmann, Digestive physiology of the Deer: Their morphophysiological specialisation and adaptation. R. Soc. N. Z. Bull. 393–407 (1985)

R.R. Hofmann, D.R.M. Stewart, Grazer or browser: a classification based on the stomach-structure and feeding habitats of East African ruminants. Mammalia **36**, 226–240 (1972)

W.L. Hurley, P.K. Theil, Perspectives on immunoglobulins in colostrum and milk. Nutrients **3**, 442–474 (2011). https://doi.org/10.3390/nu3040442

C.M. Janis, Evolution of horns in ungulates: ecology and paleoecology. Biol. Rev. **57**, 261–318 (1982).

C.M. Janis, An estimation of tooth volume and hypsodonty indices in ungulate mammals, and the correlation of these factors with dietary preference, in ed. by D.E. Russell, J.P. Santoro, D. Sigogneau-Russel, *Proceedings of the VIIth International Symposium on Dental Morphology* (1988), pp. 367–387

C.M. Janis, Tragulids as living fossils, *Living Fossils* (Springer-Verlag, New York, NY, 1984), pp. 87–94

C.M. Janis, D. Ehrhardt, Correlation of relative muzzle width and relative incisor width with dietary preference in ungulates. Zool. J. Linn. Soc. **92**, 267–284 (1988)

C.M. Janis, E. Manning, Dromomerycidae, in *Evolution of Tertiary Mammals of North America. Volume 1: Terrestrial Carnivores, Ungulates, and Ungulatelike Mammals*. (Cambridge University Press, Cambridge, UK, 1998), pp. 477–490

G. Janssens-Maenhout, M. Crippa, D. Guizzardi, M. Muntean, E. Schaaf, F. Dentener, P. Bergamaschi, V. Pagliari, J.G.J. Olivier, J.A.H.W. Peters, J.A. Aardenne, S. van, Monni, U. Doering, A.M.R. Petrescu, EDGAR v4.3.2 global atlas of the three major greenhouse gas emissions for the period 1970–2012. Earth Syst. Sci. Data Discuss. 1–55. (2017). https://doi.org/10.5194/essd-2017-79

K.A. Johnson, D.E. Johnson, Methane emissions from cattle. J. Anim. Sci. **73**, 2483–2492 (1995). https://doi.org/10.2527/1995.7382483x

W.H. Karasov, A.E. Douglas, Comparative digestive physiology. Compr. Physiol. **3**, 741–783 (2013). https://doi.org/10.1002/cphy.c110054

H. Lapierre, G.E. Lobley, Nitrogen recycling in the ruminant: A review. J. Dairy Sci. **84**, E223–E236 (2001). https://doi.org/10.3168/jds.S0022-0302(01)70222-6

G. Larson, J. Burger, A population genetics view of animal domestication. Trends Genet. **29**, 197–205 (2013). https://doi.org/10.1016/j.tig.2013.01.003

R.S. Luna, A. Duarte, F.W. Weckerly, Rumen–reticulum characteristics, scaling relationships, and ontogeny in white-tailed deer (Odocoileus virginianus). Can. J. Zool. **90**, 1351–1358 (2012). https://doi.org/10.1139/cjz-2012-0122

S.J. McNaughton, Grazing lawns – animals in herds, plant form, and coevolution. Am. Nat. **124**, 863–886 (1984).

K. Meyer, J. Hummel, M. Clauss, The relationship between forage cell wall content and voluntary food intake in mammalian herbivores. Mammal Rev. **40**, 221–245 (2010)

National Research Council, Nutrient requirements of small ruminants: sheep, goats, cervids, and new world camelids. (2006). https://doi.org/10.17226/11654

R.M. Nowak, *Walker's mammals of the world* (The Johns Hopkins University Press, Baltimore, 1999)

N.F. Owens, A.L. Goetsch, Ruminal fermentation, in *The Ruminant Animal. Digestive Physiology and Nutrition*. (Department of Animal Science, Oregon State University, Corvallis, OR 97330, USA, 1988), pp. 145–171

N.F. Owens, R. Zinn, Protein metabolism of ruminant animals, in ed. by D.C. Church, *The Ruminant Animal. Digestive Physiology and Nutrition*. (Department of Animal Science, Oregon State University, Corvallis, OR, USA, 1988), pp. 227–249

F.J. Pérez-Barbería, Scaling methane emissions in ruminants and global estimates in wild populations. Sci. Total Environ. **579**, 1572–1580 (2017). https://doi.org/10.1016/j.scitotenv.2016.11.175

F.J. Pérez-Barbería, D.A. Elston, I.J. Gordon, A.W. Illius, The evolution of phylogenetic differences in the efficiency of digestion in ruminants. Proc. R. Soc. B-Biol. Sci. **271**, 1081–1090 (2004)

F.J. Pérez-Barbería, I.J. Gordon, The functional relationship between feeding type and jaw and cranial morphology in ungulates. Oecologia **118**, 157–165 (1999)

F.J. Pérez-Barbería, I.J. Gordon, Factors affecting food comminution during chewing in ruminants: a review. Biol. J. Linn. Soc. **63**, 233–256 (1998a). https://doi.org/10.1111/j.1095-8312.1998.tb01516.x

F.J. Pérez-Barbería, I.J. Gordon, The influence of molar occlusal surface area on the voluntary intake, digestion, chewing behaviour and diet selection of red deer (_Cervus elaphus_). J. Zool. **245**, 307–316 (1998b). https://doi.org/10.1111/j.1469-7998.1998.tb00106.x

F.J. Pérez-Barbería, I.J. Gordon, A.W. Illius, Phylogenetic analysis of stomach adaptation in digestive strategies in African ruminants. Oecologia **129**, 498–508 (2001). https://doi.org/10.1007/s004420100768

F.J. Pérez-Barbería, I.J. Gordon, M. Pagel, The origins of sexual dimorphism in body size in ungulates. Evolution **56**, 1276–1285 (2002)

F.J. Pérez-Barbería, R.W. Mayes, J. Giráldez, D. Sánchez-Pérez, Ericaceous species reduce methane emissions in sheep and red deer: Respiration chamber measurements and predictions at the scale of European heathlands. Scie. Total Environ. **714**, 136738 (2020).

V. Porter, L. Alderson, S.J.G. Hall, D.P. Sponenberg, Mason's world encyclopedia of livestock breeds and breeding, vol. 2. (CABI, 2016)

R.A. Prins, R.E. Hungate, E.R. Prast, Function of the omasum in several ruminant species. Comp. Biochem. Physiol. Physiol. **43**, 155–163 (1972). https://doi.org/10.1016/0300-9629(72)90477-X

C.T. Robbins, *Wildlife feeding and nutrition* (Academic Press, San Diego, 1993)

C.T. Robbins, A.E. Hagerman, P.J. Austin, C. Mcarthur, T.A. Hanley, Variation in mammalian physiological-responses to a condensed tannin and its ecological implications. J. Mammal. **72**, 480–486 (1991)

A. Scheu, A. Powell, R. Bollongino, J.-D. Vigne, A. Tresset, C. Çakırlar, N. Benecke, J. Burger, The genetic prehistory of domesticated cattle from their origin to the spread across Europe. BMC Genet. **16**, 54 (2015). https://doi.org/10.1186/s12863-015-0203-2

R.M. Sibly, K.A. Monk, I.K. Johnson, R.C. Trout, Seasonal variation in gut morphology in wild rabbits (Oryctolagus cuniculus). J. Zool. **221**, 605–619 (1990). https://doi.org/10.1111/j.1469-7998.1990.tb04020.x

N. Solounias, M. Fortelius, P. Freeman, Molar wear rates in ruminants—a new approach. Ann. Zool. Fenn. **31**, 219–227 (1994)

E. Thenius, *Grundzüge der Faunen- und Verbreitungsgeschichte der Säugetiere. Eine historische Tiergeographie* (Urban & Fischer, Munich, Stuttgart, 1980)

C.S. Troy, D.E. MacHugh, J.F. Bailey, D.A. Magee, R.T. Loftus, P. Cunningham, A.T. Chamberlain, B.C. Sykes, D.G. Bradley, Genetic evidence for near-Eastern origins of European cattle. Nature **410**, 1088 (2001). https://doi.org/10.1038/35074088

T. Van Vuure, *Retracing the aurochs: history, morphology & ecology of an extinct wild Ox* (Pensoft Pub, Sofia, 2005)

S.E. Van Wieren, Digestive strategies in ruminants and nonruminants. Dig. Strateg. Rumin. Nonruminants 1–191 (1996)

D.J. Wuebbles, K. Hayhoe, Atmospheric methane and global change. Earth-Sci. Rev. **57**, 177–210 (2002). https://doi.org/10.1016/S0012-8252(01)00062-9

M.A. Zeder, B. Hesse, The initial domestication of goats (Capra hircus) in the Zagros mountains 10,000 years ago. Science **287**, 2254–2257 (2000). https://doi.org/10.1126/science.287.5461.2254

Chapter 3
Husbandry: Milk Production

Abdessamad Gueddari and Jesús Canales Vázquez

Abstract This chapter briefly describes the production systems of dairy farms, and in addition, the most conventionally used milking processes and operating models are discussed. After that, the constructive elements and the required facilities for the development of dairy farm activity are explained. Finally, the management and treatment processes of the waste generated in these farms are presented, pointing out the best available techniques and their most commonly used methodologies to mitigate environmental impact, with a perspective of the new most promising emergent reusing dairy wastes.

3.1 Introduction to the Dairy Production Systems

A production system can be defined as the combination of a series of resources, techniques and operations to achieve an adequate coordinated management over the dairy farm production parameters (land use, capital, production capacity, animal welfare, milk quality, labour use, etc.). This enables the outcome of best quality products plus the optimization of the use of resources over time. Thus, a farm works in a sustainable manner, ensuring also the animal welfare, as both these aspects have become a major concern in food industry over the last decades, especially in developed countries.

According to FAO, there are more than 278 million dairy cows worldwide. This means that over 602 tonnes of milk are produced annually worldwide considering an average annual yield milk of 2200 kg per cow. India, Brazil, Pakistan and China have the largest inventory of dairy cattle with 51, 17, 13 and 12 million cows, respectively, followed closely by Ethiopia and USA with 12 and 10 million cows. In any case, there are also some other figures which must be considered as more cows does not necessarily imply larger productions. Indeed, USA exhibits the largest milk yield worldwide with almost 88 million tonnes of milk produced annually (FAO 2006), and this is a clear example of wide variety of production systems in dairy industry (FAO 2006). Milk is not produced by a unique standardised dairy production system due to a combination of parameters, such as demand, technology evolution, geoclimatic conditions and cultural issues. Considering that worldwide milk production is

© The Author(s), under exclusive license to Springer Nature Switzerland AG 2020
S. García-Yuste, *Sustainable and Environmentally Friendly Dairy Farms*,
SpringerBriefs in Applied Sciences and Technology,
https://doi.org/10.1007/978-3-030-46060-0_3

projected to increase up to 1043 million tonnes by 2050 (FAO 2006), analysing the
parameters that may determine milk yield becomes a major issue.

Nowadays, there are several types of dairy cattle systems since each production
model exhibits a unique complex combination of circumstances, and therefore, there
are multiple possible ways to classify them. Nevertheless, a high-level classification
is usually carried out according to the feed management nature, livestock management process and characteristics of the erected facilities. According to the housing
conditions, there are three main groups of production systems: (1) extensive (pasture-
based); (2) intensive (confinement); and (3) mixed (semi-intensive). Most dairy cows
in Europe and North America are based on confinement systems. In contrast, extensive systems are mostly raised in South America, Australia, India and New Zealand
(Endres and Schwartzkopf-Genswein 2018). Figure 3.1 shows the flowchart diagram
of a typical dairy farm. The following sections describe each of these dairy systems.

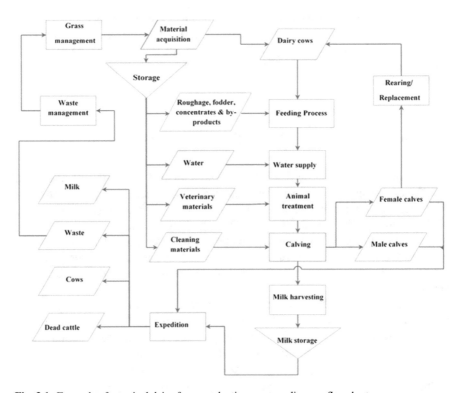

Fig. 3.1 Example of a typical dairy farm production process diagram flowchart

3.1.1 Extensive Production Systems

This model emerges from the first total grazing dairy farms developed by humankind, but in this case small resources input are exploited. The extensive production systems are farm models where dairy cows satisfy their food needs in a traditional way, that is, they consume the natural resources provided by the farm environment; in other words, they are free range for part or all their production cycle. Thus, it can be considered as an imitation of the natural ecosystems, which implies the utilisation of large (extensive) grazing areas.

The production cost of this system is usually lower than that of the confinement systems as long as the farmer is the owner of the grazing land or in the case of inexpensive renting land. Since the species are not specifically selected, autochthonous livestock is acquired from the same area. On the other hand, feeding comes from the natural environment, which means that there is an important seasonality of the pastures and fodder availability. Therefore, a greater management and control of the stocking density must be implemented. This would ensure a greater ingestion and pasture use by the animal and therefore, optimal conditions will be reached involving milk production and quality maximisation. The advantages of this model include the following:

(a) The autochthonous cattle exploitation contributes to the consumption of the unproductive natural resources, leading to maintain the biodiversity and the natural agroecosystems stability.
(b) This model operates with a lower carbon footprint compared to other systems, given that lower operating energy is consumed per unit of product obtained.
(c) Good animal welfare due to free range farming method that contributes to a more favourable animal growth as long as cows roam freely outdoors. This has become a major issue in developed countries as end consumers have become very concerned regarding animal exploitation.
(d) Independence from suppliers of agricultural products.

On the contrary, these systems have the following disadvantages:

(a) The production efficiency per unit area is lower than the other models added to a very high seasonality.
(b) This model involves late production cycles since the animal productivity depends on the grazing management and the forage accumulation, which in turn relies on several parameters such as climate and land conditions. For this reason, in some cases it is necessary to move searching for food, delaying the production cycles. This delay involves difficulties in adjusting to consumer demand.
(c) Providing homogeneous products within the criteria established by the market is a highly demanding task, especially for this kind of production system. This is due to the lack of products standardisation coming from heterogeneous pastures and cows reproduced in indiscriminate groups.

3.1.2 Intensive Production Systems

Intensive production systems are based on a very standardised exploitation methodology based on zero-grazing dairy farms, in which animals move in a limited space. They satisfy their feeding needs inside their own confined animal stalls, receiving food inside the facilities using feeders, under optimum conditions that seek to maximise the milk production performance. As can be expected, confined production systems have been developed, thanks to technological and scientific advances related to the following key issues: animal feeding, hygiene, welfare, dairy tools and genetics. Since farms with such production systems incur high operation costs (compared to extensive systems), it is necessary to combine all these factors adequately to ensure profitability and stability over time. For instance, the animal feeding must be totally under control, adjusting mixed rations with forages and concentrates which are supplied in due amount and time leading to a marked increase in milk production. As a consequence, each cow in intensive systems produces between 7,000 and 14,000 kg during 305 days of lactation while cows in extensive systems produce between 3,000 and 7,000 kg per cow during the same period (Endres and Schwartzkopf-Genswein 2018). Generally, the incurred costs associated to this production system are 60% as consequence of feeding costs, 18% due to inversions, 15% because of labour resources and other 7% due to other expenses such as cleaning product and veterinary.

Unlike extensive production systems, these systems allow greater control of the feeding quality and quantity consumed by cattle, providing a stable production cycle with balanced diets and continuous supplies. Furthermore, the land is usually adopted according to the number of the dairy cows per farm, being the agricultural activity and livestock independent. Nevertheless, selective cattle are chosen to ensure reaching very high production ratios (Blount 1968).

Buildings and facilities play a crucial role in the maximisation of milk production in these systems. They must be projected and designed properly to guarantee suitable housing conditions, and therefore to maximise production, parameters such as ventilation, humidity, temperature and hygiene require an utterly strict control. This allows the maximum welfare of the cows and therefore better global process performance.

From the point of view of workforce as more complex advanced techniques are used, a more skilled labour would be needed. Moreover, these systems, as can be expected, require not only plant-level workforce but meticulous management as an industrial business unit. In other words, in advanced production systems, there is a growing number of highly prepared professionals fully aware of the relevance of all the parameters mentioned above and how necessary they are to control all the aspects involved directly or indirectly in the milking process.

Finally, it should be noted that some of the current intensive production systems are moving towards new changes, thus engendering a new variant called industrial exploitation system. This new system could even be assumed as a new methodology of livestock model, which is based on intensified factory farms based on larger intensive units. The main characteristic of these systems is feeding, as specific indus-trial products are acquired exclusively for some species of cows that do not require feeding coming from pastures.

3.1.3 Mixed Production Systems

Another production system that is usually implemented is the mixed type, which con-sists of mixing the methodologies and management techniques of both the systems described above. These systems are very variable, as the characteristics of both oper-ating systems can be combined in several ways. This model combines the advantages of both intensive and extensive systems, besides allowing efficient nutrition control of cows to improve their yields, depending on the batch of species of the cattle. Finally, it prevents cows from being subjected to adverse climate as would occur in the extensive system.

The main objective of this mixed system is to cover the increased milk demand, reduce cost by labour-saving technology and high efficiency by a suitable grassland managing and using of the new technology. On the other hand, EU has implemented policy changes oriented to stimulate the use of grass again as an environmentally friendly production which reinforces the future need of combining both systems (Kristensen et al. 2005).

3.2 Milking Systems

Milking routines involve more than 50% of the total labour needed to carry out the global process (Andrews et al. 2016). This means that an efficient dairy milking pro-cess management is required, projecting the most suitable milking centre depending on all the involved factors. There are several types of milking systems, which can be classified into manual and mechanical technologies, depending on their level of automation. On most occasions, small farms tend to erect manual milking systems, while larger production farms implement mechanical processes. This is due to the workload involved in this milking process, as milking usually takes place from two to three daily shifts, which implies most of the production cycle time. Under these circumstances, this process phase is decisive, which is why not only routine control is required but also adequate infrastructures that guarantee maximum milk production with the best possible quality.

Manual production systems mean a lower investment cost for the farm with no over-milking risks. But, conversely, this means lower production efficiency due to the manual process, with lower hygienic quality and worse working conditions compared with those provided by mechanical system. Consequently, this is translated into higher overall milk production costs when a high level of milk production is available.

This section describes the mechanical advanced milk production types, as they are globally most common.

Mechanical milking systems can be generally classified into two milking models depending on whether it is open or fixed housing conditions. In the fixed housing model, the milking process is carried out in the barn, which means in the same place where they get fed (stall milking). In the other case, an independent milking parlour is used (parlour milking). Open housing is an organisation adopted to ensure animals can move freely in the stable premises, either outdoors or indoors. Otherwise, the cows have limited movement in the fixed housing method. Nevertheless, it must be highlighted that in recent times, a lot of dairy farms are including exercise areas to improve cow health, since many studies demonstrate the positive effects on the cows health due to daily exercise, reducing the need for veterinary treatments (Gustafson 1993).

The tied stalls are based on the organisation where the female cattle are usually milked in their housing place by extracting the milk accumulated in their udders through the milking unit. Conversely, in the case of milking parlour, it is the cattle that move towards farm's parlour milking area in order to develop the milking process. The functions of milking parlour are to improve and facilitate the process of obtaining milk, as it not only improves the overall quality of milking but also allows greater animal comfort. Additionally, it reduces infection risks as they ensure better hygiene conditions, since the milking area is usually independent with respect to the other barn dependencies. This distribution also maximises the overall process productivity plus making way easier the milker. In many countries with advanced milking processes, there are state regulations that require these milking parlours dedicated exclusively to the milking processes to meet these objectives. These two milking methodologies are described and compared in the following section.

It should be noted that in recent times an alternative fixed housing has been adopted in which an independent milking parlour is available as a solution, allowing the animals movement through the farm exercising their muscles, which contributes to prolong their lifespan, as well as reduce labour and thus facilitating the overall management of the holding.

3.2.1 Tied-Stall Milking Systems

Within the tied dairy milking systems, there are two different work typologies: churn milking and round-the-shed milking.

(a) Churn milking

This milking methodology is mainly used in small herds with low investment capacity. The milk is extracted into a mobile container (milk churn) connected to the vacuum system and milk is produced by the pump located outside the stable. Once the container is filled, it is transported to the milk cooling tank.

Since it is a portable and small system, its investment cost is low, with low maintenance costs and very easy to use. Otherwise, this system has the disadvantage that the milker performs all operations manually, reducing the overall productive efficiency of the production system. Also, during manual transport of the milk to the storage tank, there is a high risk of milk contamination.

(b) **Round-the-shed milking**

In this case, as in all systems, milk is extracted by means of the milking unit, but instead of being deposited in a churn as in the previous case, it flows into a common vacuum pipe, running through the whole stable and discharging into the final milk storage unit. This working methodology is most often used for a moderate number of cows and is especially used in areas with very cold weather conditions (Central European and Scandinavian countries). The drawback is the poor optimisation of the workforce in poor ergonomic conditions plus the poor milk treatment that must travel long distances to reach the points of collection to be stored.

3.2.2 Parlour Milking

In this model, a new area called milking centre is built totally independent of the barn destinated solely for milk production. This milking centre consists of the parlour, a holding area, utility room and other optional areas like the supply room intended to develop the milking process. The holding area is where the animals are concentrated to pass from the stable to the milking parlour. This holding area must be designed to tightly fit all cows since all of them are brought to the parlour in group (Endres and Schwartzkopf-Genswein 2018). In most cases, the utility room includes ancillary facilities to manage the produced milk like the refrigerated milk tank for storage and the engine room where the pump-motor set is located.

There are several milking parlour configurations depending on the milking routine, production level as well as the available space area and workforce. These types are usually classified into parlours arranged in series or in parallel, depending on the adopted cattle direction during the milking process. The series milking parlours are designed to ensure that cattle entering and leaving the milking process have the same direction. On the contrary, cows entering and leaving the milking process in parallel systems are usually parallel to each other in a perpendicular direction to the milking pit. Serial systems are in turn classified into two possible very common variants called tunnel and tandem. Alternatively, parallel systems are classified into classic parallel (flat barn), herringbone and swing parlour. Furthermore, all these variants can be classified into single or double typology, depending on whether the cows are moved on a single platform as corridor or if they are located on two platforms parallel to the milking pit, respectively (Buxadé Carbó 1995). Each of these solutions is described below.

(a) **Tunnel parlour**

In these parlours, cows intended to be milked occupy the entrance/exit lanes. The cattle enter the corridor in line, occupying the free stalls to being milked and then continue their way to the exit through the same lane.

(b) **Flat-barn parlour**

In this configuration, the parlour access from the holding area to the independent stalls is designed in a perpendicular direction to milking pit main axes. Cows normally access by a single step-up, allowing an easy access to the udders from the rear. The cows move forward towards the return lane. This is the low-cost milking parlour configuration, though this configuration is not as labour-efficient or ergonomic as the elevated parlours (Reinemann and Rasmussen 2011) (Fig. 3.2).

(c) **erringbone parlour (fishbone)**

Cows stand in an elevated platform slanted towards the major axis of the pit, with the udders close to the pit, facing away the milker area at an angle of about 45° (Reinemann and Rasmussen 2011). Cockburn et al. (2017) studied the effect of other fishbone angles, recommending finally 30° as the best position to improve the personnel comfort. This herringbone configuration allows a greater flow of cows into the system. In this case, unlike the previous described, the entry and exit of the cows is performed in batches, being the time of the cycle defined by the cows requiring longest to be milked (bottleneck). With this methodology, a very high output of up to 100 cows per operator per hour may be achieved, requiring a smaller surface area compared to other typologies. The disadvantage of these systems is that it is not possible to treat cows individually, as treatment is carried out in batches.

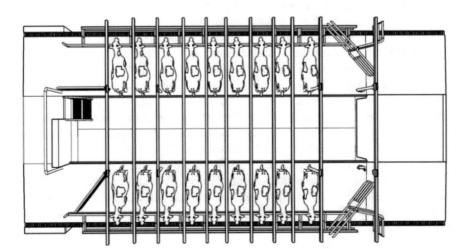

Fig. 3.2 Flat-barn parlour configuration

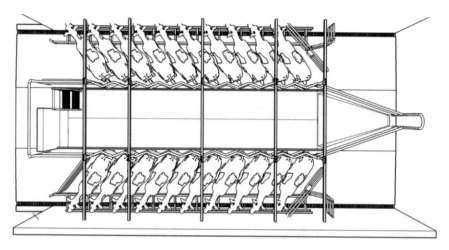

Fig. 3.3 Herringbone parlour configuration

With the aim of optimising the production cycle times, several variants of this parlour type have emerged, such as triangular and polygonal configurations, where the animals are subdivided into several batches, but following the same organisation. This parlour configuration is one of the most common built around the world (Kutz 2007) (Fig. 3.3).

(d) **Side open parlour (Tandem)**

In this case, unlike tunnel parlour configurations, the entrance/exit lanes are independent from the milking stalls. These stalls are usually parallelly directed to these lanes. Cows stand in an end-to-end arrangement over the milking process (Reinemann and Rasmussen 2011). This organisation allows cows to be milked from the central pit side, which is at least 0.9 m deep for the operator's sake avoiding any possible accidents due to cow kicks. Furthermore, this design facilitates access to the udder promoting the possibility of automation to ensure higher yields. Finally, these systems require large surfaces and therefore the operator must travel long distances wasting more time than in other compact configurations (Fig. 3.4).

(e) **Swing over parlour**

This configuration can be considered as an improved version of herringbone parlours, where the difference lies in the standing angle. In this case, cows form angles in the 70–90° range with respect to the milking pit, improving the movement of the animals. However, this makes the process more difficult for the milker to see the udders due to the angle visibility reduction. The main advantage of this layout is that each milking unit is placed on two milking parlour sides. Therefore, this configuration entails milking fewer cows per stall but more cows per milking unit over the same period of time compared to parlours with just one milking unit per stall.

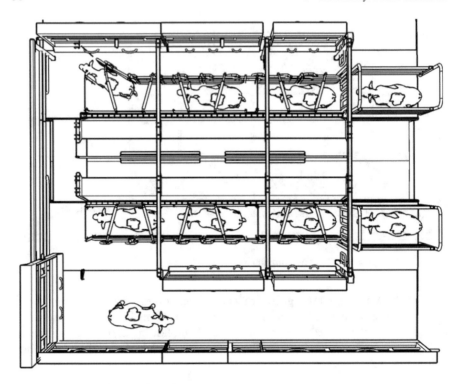

Fig. 3.4   Tandem parlour configuration

During the last several decades, new solutions have emerged with other arrangements for large herds, such as movable and rotary parlour.

(f)   **Movable parlour**

These systems consist of the automation of the process, where the cows are moved linearly by mechanical systems. These systems are underused due to the complexity of the process design and the high implementation cost.

(g)   **Rotary parlour (carousel, turnstile)**

The rotary parlours are a very common solution where the dairy cows stand on a rotating platform. This configuration avoids the milker displacements to access the cow udders, that is, he/she remains in a fixed position. This layout increases the milking process performance, as the time required to connect the milking units to the udder is reduced, with a continuous production plus optimising labour and ergonomic conditions. Due to automation and movable mechanic power, the capital cost is usually higher in comparison with non-moving parlours. Therefore, they are

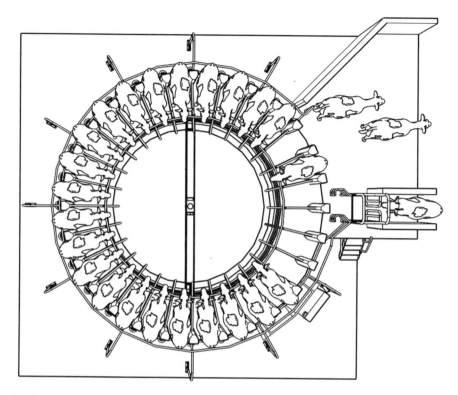

Fig. 3.5 Rotary parlour design

suited to a minimum of 1000 cows. The number of stalls in these systems can vary from 10 to 60 or more (Kutz 2007).

There are several layout designing parameters depending on the position of the cattle and the operator, which can be classified according to the 'face-in' and the 'face-out' arrangement. The last is the most frequently used because the cows are in a roto-radial direction and the operator is outside the platform, controlling visually the parlour efficiently. Otherwise, the interior milking configurations are those where the milker is in the middle of the platform, and cows in a similar fashion as the herringbone parlour (roto-spine) or tandem (roto-tandem), but radially (Buxadé Carbó 1995) (Fig. 3.5).

3.2.3 Automated Milking Systems

During the last decade of the past century, the first automated milking systems (robotic milking) were introduced in the Netherlands. In 2008, around 5,500 farms adopted this technology. In these systems the cows are milked 2–3 times per day without the support of any operator. The main objective of this new system is to change not only the technology but also the whole management system, resulting in increased flexibility and milk production and decreased workforce need up to 15 and 18%, respectively (Svennersten-Sjaunja and Pettersson 2008; Andrews et al. 2016). But instead, the main disadvantages of these systems are the high investment and maintenance costs (Hansen 2015). This new model affects negligibly the milk quality and properties increasing the global farm efficiency (De Marchi et al. 2017).

These systems are also called voluntary milking systems since cows decide her milking frequency during the day. Cows always try to enter to the milking machine to reach the tastiest feed. When the cow decides to enter, the control system detects her identification to decide if it is suitable to be milked on the basis of her historical milking status. The cycle consists of several steps: udder cleaning, milking cup fitting, tactile stimulation, milk extraction, milking cup unfitting and finally equipment cleaning. This automated system includes electrical conductivity to detect contaminated milk coming from problematic cows to store it in an independent tank (Andrews et al. 2016).

Automated milking systems are divided into single stall and multi-stall systems. In the single stall the robot could be used for three or four stalls with 60 cows, demanding an independent cleaning device. However, the robot in multi-stall systems can serve 160 cows (Andrews et al. 2016). Milking labour cost reduction in the case up to 31% per year has been estimated in the case of using robotic milking system instead of conventional ones (Hyde and Engel 2002).

3.3 Farm Construction Elements

Many factors should be considered when planning a building to obtain great benefit in order to guarantee cows welfare at a proper cost. These factors are related to the functional, structural and material designing, where space, temperature, light, physical protection, insulation, value, durability, sanitation and other factors must be considered (Andrews et al. 2016). In the following section most important farm construction criteria are discussed.

3.3.1 Enclosures and Roofs

The enclosures and roofs of dairy farms are usually designed according to the climatic conditions of each area, as it must allow for correct isolation depending on cattle species housed, as well as other hygiene and safety factors.

The lateral enclosures can be made of galvanised sheet, in situ working concrete or pre-fabricated. These enclosures not only isolate the building from adverse external environment but also prevent cattle from escaping. There are many solutions to improve the insulation in order to improve cows' welfare, such as sandwich panels, plastic dropped ceilings, carbon fibre panels and projected polyurethane. Furthermore, interior coatings are often used to protect buildings against corrosion due to possible condensation or dripping. In addition, this coating facilitates the cleaning of the building elements indeed. With regard to enclosures and roofs, external coating is usually applied in order to minimise the visual impact of the installations and to improve element protection against corrosion. The most common coverings are plastered and plastic paints. In the case of interior coverings, for the roof polyurethane is usually used, while on the outside it is usually painted or lacquered. In the same way, tiling is normally used, and although it is more expensive, in the long term it is usually profitable because of its low maintenance cost. Finally, in the case of milking parlours, their enclosures are usually carefully designed to guarantee an efficient isolation from contamination sources, with easy-clean surfaces to enhance the hygienic conditions (Endres and Schwartzkopf-Genswein 2018).

3.3.2 Flooring Considerations

Floor is a constructive element with which the animals are in continuous contact, hence it is necessary to choose and design it to give the greatest cattle comfort. In addition to withstanding pressure loads, the function of the floor screeds is to guarantee thermal insulation and absorb humidity to avoid affecting cattle welfare. In many cases, some auxiliary elements are installed, like organic and inorganic cow bedding or advanced solutions like waterbeds depending on the farm area (Andrews et al. 2016). The following are the most common design criteria for the important farm areas.

(a) **Exercise Area**

Exercise areas are uncovered areas where cows can have small space to move around, thus improving its welfare when it comes to fixed stalls. The floor must be easy-clean, with a slight slope towards the perimeter socket of the slurry pit. In this area the floor should be made of dirt or concrete. Dirt floors are cheaper than concrete floor, but are harder to clean, which increases the operation cost. The concrete floors, added to their easy and fast cleaning, represent better access for the cleaning machinery. When cleaning operations are carried out mechanically, paved floor screeds are usually chosen, as they are more resistant and stable against machine loads, and do not require periodic levelling.

   In this area, cows must have a minimum space equal to the useful area of the resting area, as it must ensure good circulation of animals, people and cleaning machines. The exercise area required for dairy cows is approximately 6–7 m^2 per cow (Kutz 2007). Given the surface area requirement, in some cases, when the areas are very large, the tendency is to pave only the areas adjacent to the feeders and the stable to reduce costs.

(b) **Cattle shed area**

Accommodation areas are usually designed according to the model of exploitation system adopted, as well as species, climate and economic factors. The floors of this housing area are usually made of concrete in order to improve dragging of the droppings (well-drained floor) and also the cattle housing conditions. Bedding systems are always adopted to act as an insulator between the cattle and the ground, making the resting and calving areas more comfortable plus minimising the disease risks. Organic materials, like straw, rice hulls or sawdust, support better bacterial population than the inorganic materials such as washed sand and limestone (Andrews et al. 2016).

(c) **Feeding area**

The floor in the feeding areas is usually made of concrete coated with non-slip materials. Some feeding areas include also bedding materials, which allow a high degree of comfort to the livestock during the cows feeding. There are other screeds that tend to be more comfortable for cows, which are those of rammed earth, but in return this implies a high maintenance cost. It is also common to install graded-aggregate floor or cubicles covered with non-slip and hygroscopic carpets.

(d) **Waste management area**

The screeds of the pits and dunghill must be paved and correctly waterproofed to avoid contamination by leachate. The most commonly used cleaning system is the scraper, which consists of a hydraulic drag system that removes the droppings from the corridors to the accumulation pits. The other existing technologies for this propose are flushing and vacuum track. On the other hand, the carcasses installed for the management of waste produced on the farm are usually covered with iron grids to prevent cow injuries. Nowadays, pit flooring is designed to guarantee some flexibility with interlocking comfort tiles to minimise backaches and leg cramps (Andrews et al. 2016).

3.4   Environmental Control

In order to ensure animals welfare and improve animal production as well as worker comfort and facilities durability, appropriate environmental control is required.

3.4.1   Heat Stress

Dairy cows are homeothermic and sensitive animals which require an environmental control of different factors to avoid influence on their thermal exchange and therefore their comfort (Andrews et al. 2016). The most important factors are solar radiation, air temperature, air movement and humidity. There are several strategies to prevent heat stress focused on controlling these four main factors added to feeding time cycles and crossbreeding. Some strategies to reduce solar radiation consist basically of increasing the shaded areas, typically 4–5 m^2 per cow. In the case of controlling the air temperature, air movement and humidity, ventilation and cooling systems must be properly projected. Cows cool themselves by evaporation, conduction, convection and radiation heat transfer, but if they are exposed to uncontrolled environments with an impact on their thermal exchange, metabolic processes, cow behaviour, milk yield and milk composition will be directly affected, hence increasing the risk of disease.

In general, cows are in comfort when the ambient temperature is between 5 and 20 °C and the temperature–humidity index (THI) is up to 68–72% in addition to adequate ventilation systems. THI is a quantitative number that combines temperature and humidity impact on cows. THI values between 72 and 78% mean that cows are in the mild stress zone, while values between 79 and 89% mean they are moderately stressed, and THIs between 80 and 98% mean they are in the severe stress zone. Cows die because of heat stroke when THI is above 98% (Armstrong 1994). Consequently, THI must be monitored and modified conveniently to avoid such dangerous scenarios.

3.4.2 Ventilation and Air Quality

A mandatory/imperative process of confinement housing is the continuous indoor air exchange with outdoor fresh air in order to maintain indoor safe conditions. Ventilation is based on air renewal which maintains harmful gases under safe levels ensuring animal and human welfare. Air renewal consists of removing dust and harmful gases (CO_2, NH_3, CO, CH_4, NO_2, H_2S and water vapour), which are produced by the dairy activity and via respiration and faeces decomposition that directly affect the indoor air quality, as some of these gases are harmful, and tend to generate odours affecting the humidity and environment oxygen concentration. Therefore, with an appropriate air renovation, the quality of the indoor air is improved, as well as the physiological performance of the animals, thus reducing the risk of suffering from diseases such as mastitis, pneumonia or diarrhoea. Table 3.1 shows the gas maximum permissible concentration levels in dairy buildings and their class effect (Andrews et al. 2016).

Table 3.1 Maximum concentration of most important gases in dairy facilities

Gas	Class	Concentration limits (ppm)
Ammonia	Irritant	20
Carbon dioxide	Asphyxiant	3000
Carbon monoxide	Poison	10
Hydrogen sulphide	Poison	0.5
Methane	Asphyxiant, flammable	117

Ventilation also allows better temperature and humidity control, improving animal and human comfort, as well as reducing the pathogenic micro-organism concentration and facilities corrosion and degradation risks (Andrews et al. 2016). On the other hand, it should be noted that, in any workplace, the work area must have a minimum air quality required to avoid health risks to operators. Ventilation systems must be able to guarantee sufficient fresh air exchange, the minimum value being 4 air change per hour in the winter due to low temperatures and a maximum value between 40 and 60 in the summer with an optimum air flow circulation speed of between 0.85 and 2 m/s (Mondaca 2019). There are different ventilation system types to carry out these air renovations, which can be divided into three technologies: static (natural), dynamic (mechanical) and the combination of both systems (hybrid barns). Each of these possible typologies is explained below.

(a) **Static ventilation**

In this system, indoor air is renewed via the chimney effect. This effect consists of the movement of air due to the density gradient caused by the temperature gradient that causes a pressure difference between the indoor and outdoor air known as thermal thrust. In order to guarantee this natural ventilation, buildings are designed in proper dimensions, with openings and orientation of their construction elements with respect to the winds in such a way so as to guarantee thermal thrust. When fresh air enters the farm, it is heated by mixing with the interior air, which leads to a density drop that makes its rise to be dislodged by chimneys or openings intended for that purpose. At the outlet, it creates a certain depression that favours the entrance of more external air and thus cyclically.

(b) **Dynamic ventilation**

In this case, fresh air renewing is carried out by exhaust and intake fans. When dynamic ventilation is chosen no orientation and meticulous architectural designs are necessary, and moreover, the requirements regarding ceiling and placing distances are not so strong. Mechanical ventilation is typically used in parlours and milk rooms (Reinemann and Rasmussen 2011). Some disadvantages of these systems are the electricity consumption, maintenance cost and the need of a backup generator in case of emergency. There are three types of ventilation systems: positive, neutral and negative pressure. Positive pressure systems consist of fans blowing air into the building creating the sufficient positive indoor pressure to exhaust air through projected openings. In contrast, negative pressure systems consist of fans expelling indoor air, creating a negative indoor pressure with respect to outside. Finally, neutral pressure systems are based on maintaining an equilibrium between indoor and outdoor pressures. On the other hand, there are two types of mechanical ventilation configurations in use in dairy industry: tunnel and cross ventilation. The first shows air flows parallel to the feed lane, whilst in the second the air flow is perpendicular. Cross-ventilation is only the choice in large housed systems due to high costs (Mondaca 2019).

(c) **Hybrid barns**

A combination of natural and mechanical ventilation is implemented, in general, in a tunnel configuration with adjustable ridge or ridge cupola fans. This system is based on controlling the environment during all seasons, where mechanical ventilation with circulation fans works during the summertime to ventilate the barn, hence reduces thermal load, whereas only natural ventilation is used in winter. When THI drops, the ventilation rate is reduced automatically in order to reach the set-point. Usually, the THI set-point is around 68, and below this value no fans are needed. This system, in contrast with mechanical ventilation, allows to renew the air in case of an emergency. Other benefits associated to this system are the reduction of heat stress and prevention of manure getting frozen during winter. This type of ventilation is particularly interesting in regions exhibiting large temperature variations (Mondaca 2019).

3.4.3 Heating and Cooling

The metabolic heat produced is directly related to milk yield: higher milk yields imply higher heat load, which can be reduced by wetting the animal skin (Flamenbaum et al. 1986). At ambient temperature, heat is mostly dissipated via evaporation (Flamenbaum et al. 1986). The most commonly used evaporative cooling systems are based on sprinkling or dripping water at a rate of $0.2 \ 1 \ m^{-2} \ min^{-1}$ on cattle in the feeding area. Evaporative cooling is efficient in dry environments and when the air speed is between 1 and 1.5 m/s (Flamenbaum et al. 1986). On the other hand, it should be noted that evaporative cooling systems by wetpads demand intensive maintenance tasks due to the high risk of clog-up with water residues (Mondaca 2019).

In cold areas, in general, it is rather common to set up a heating system in the milking parlour and offices and have separate ventilation systems in the milking parlour for cold weather to remove moisture and warm up (Reinemann and Rasmussen 2011).

3.5   Waste Management

The waste generated by dairy farm activities can be classified as inorganic and organic residues. The generated amount of wastes is usually highly variable as it generally depends on the adopted livestock model, facilities, feeding system and cleaning program (Brownlie and Henderson 1984). These two types of residues are described below.

(a) **Inorganic wastes**

Inorganic wastes are all those wastes of non-biological nature that are generated at the livestock farm and cannot be composted. These are usually comprised, on one hand, of the wrappers coming from the packaging of feed and straw. Most of these residues are usually managed by the supplier companies themselves or by urban management companies after their accumulation in areas set aside for this purpose. On the other hand, zoosanitary products and biocides are generated by veterinary processes, such as needles, medicines and packaging. Most of them are considered dangerous due to their toxicity, so they must be disposed through a selective management system to guarantee their correct treatment.

Finally, cleaners are produced from farm facilities and equipment washing processes. Polluted waters arise from bulk milk tank rooms, milking parlour pits, milking parlour cow standing areas, outdoor fouled yards and silage clamps are typically diverted away from slurry draining system and collected separately (Brewer et al. 1999). Dairy parlour washings have been estimated at 50 l per cow per day (Rovira et al. 1996). This sewage tipping may need special treatment if milk solids and cleaners concentration are high due to their difficulty to be degraded (Andrews et al. 2016). The most common treatment systems are usually septic tanks. In these pits the solid matter is decanted to the bottom of the container. Since it is a closed system, the residue is fermented anaerobically, consuming part of the residue by microbial metabolism. Some studies show an efficiency above 80 and 90% of removal dairy parlour wastewaters and suspended solids, respectively (Luostarinen and Rintala 2005). Prior to the discharge of these graves to dirty water system drainage, an authorisation is usually requested from an authorised control of pollution manager.

(b) **Organic wastes**

Organic waste involves any compound coming from animals or vegetables. In these farms, various types of organic waste are generated. The residues with the highest level of production are faecal water and animal faeces (manure). In addition, other waste is generated such as dead animals, waste from calving or milk remains during cleaning. Manure properties depend on a lot of circumstances such as type of housing system, animal species, diet and environment. Manure quantity and composition determine the whole livestock design. Representative values and characteristics of the daily dairy excreted manure according to (ASABE 2005) are shown in Table 3.2.

It should be noted that Table 3.2 reflects 'as excreted' manure characteristic, which is different from the recollected dairy manure effluents since each farm uses specific

Table 3.2 Estimated typical manure (urine and faeces combined) characteristics ('as excreted')

Dairy production grouping	Solids	Chemical oxygen demand	Biochemical oxygen demand	N	P	K	Total manure	Moisture
	kg/day · animal (d · a)						kg/(d · a)	%w.b.
Lactating cow	8.9	8.1	1.3	0.45	0.078	0.103	68	87
Dry cow	4.9	4.4	0.63	0.23	0.03	0.148	38	87
Milk fed calves				0.0079				
Calf-150 kg	1.4			0.063			8.5	83
Heifer-440 kg	3.7	3.4	0.54	0.12	0.02		22	83
Veal-118 kg	0.12			0.015	0.0045	0.0199	3.5	96

methods and techniques to manage the generated manure as explained in the previous sections.

Manure can be classified according to composition and solid fraction. On the one hand, a diluted liquid mixture with a solid fraction up to 4% is generated and collected in the liquid lagoon in order to be directly irrigated. On the other hand, a mixture with a solid fraction between 4 and 12% is naturally generated and handled as a slurry with special pumps. Conversely, manure usually contains solid fractions above 20% because it is a mixture of faeces excreted with urine and plant material such as feedstock, and other materials such as straw or wood shavings that can be dragged from the stables in which it is usually used as animal bed. This manure is handled as solid by bucket loaders or forks. Intermediate residues containing between 12 and 20% of solids are considered semi-solids, which are very difficult to manage, and in general, water draining is normally used to handle manure as liquid (Lorimor et al. 2008).

3.5.1 Manure Management

Farms produce large volumes of slurry and manure which require an adequate management program to protect the environment and to produce not only in an eco-friendly way but also efficiently (IAEA 2008). There are several management systems for livestock waste, which can be classified as conventional and advanced. Conventional systems consist of natural models that are based on the direct use of these wastes in soils with fertilizer needs. Alternatively, advanced systems use complex techniques and methodologies for the treatment of these produced wastes prior to their final use in soils.

These latter techniques are increasingly used due to the growing evolution of livestock farming systems, which have evolved towards intensive systems with high density of livestock in the last past two decades. Such trend can be considered as a driving force to adopt alternative management models of livestock dropping towards more sustainable systems and avoiding negative environmental impact. Consequently, advanced waste management techniques are currently being implemented to control and prevent nitrate pollution of soils in vulnerable areas. In addition to soil pollution, these systems are intended to comply with the Kyoto protocol and Gothenburg as far as N_2O, CH_4 and NH_3 emissions are concerned. In addition to other legislations, Directive (1991) is applied in Europe to protect the water pollution from nitrates NO_3^- (Chadwick et al., 2011). The most commonly used waste management techniques are described and characterised as follows.

3.5.1.1 Direct Use of Slurry

The direct use of slurry on the soil surface without pre-treatment destined to fertilize farmland is the cheapest and most conventional technique, where farm resources are exploited efficiently. However, given the variability of the nutrient composition of the slurry, it is necessary to analyse them for a more controlled application that would prevent negative environmental impact. Nevertheless, it is also necessary that the land and agriculture activities are fully integrated to ensure the applicability and coordination of this technique. To date, the application of this technique also depends on the soil type and its cultivation, being only applicable in areas where there are no limitations of nitrogen and phosphorus concentrations exceeds in soils. After all, these techniques entail also atmospheric emissions which must be controlled. The environmental risks derived from this manure management methodology are as follows (Chadwick et al. 2011):

(a) Emission of greenhouse gases: nitrous oxide, N_2O, carbon dioxide, CO_2, and methane, CH_4.
(b) Ammonia emission, NH_3.
(c) Eutrophication of water due to soil contamination.
(d) Bad odour emissions.
(e) Damage to the ecosystem and health risks.

Usually, these risks are partially mitigated by reducing their environmental impacts through the scrupulous application of best available techniques (BAT). Alternatively, when this methodology cannot be applied due to terrain conditions, other strategies such as the transport of slurry or its storage may alternatively be adopted.

(a) **Slurry transport**

When direct slurry irrigation in nearby areas is limited by terrain conditions or its vulnerability to contaminants, the transport of slurry by watertight cisterns for application to non-vulnerable areas is often used. This solution is economically limited by the distance to be covered due to fuel consumption. The average transport cost of slurry is around 0.15 \$/km m³.

(b) **Slurry storage**

This alternative consists of accumulating the slurry in storage systems for subsequent application depending on the agronomic need of the land. There are several types of slurry storage systems. The most common storage systems are external tanks, ponds, lagoons and bladder tanks. In general, the decision of using one storage system or another frequently depends on the storage system cost and farm characteristics. However, the chosen system must avoid the emissions of bad smells and particularly ammonia. It is therefore common practice to reduce the contact area of the stored slurry free surface with the air to prevent methane emissions. Another possibility might be the use of open systems, but it is necessary to resort to capping, without having an anaerobic system that can promote fermentation by generating other gases. Finally, it is important to remark that the agitation of these stores is generally avoided, and the containers are properly insulated to guarantee water tightness and avoid leak risk (Bittman et al. 2014).

3.5.1.2 Pre-treatment Techniques

When slurry cannot be applied directly to land because of regulations or other restrictions due to nutrient excess concentration, pre-treatment techniques can be used as a solution. The solid–liquid separation technique is one of the most used process where liquid and solid phases are segregated recovering nutrients from slurry. It is worth noting that neither organic matter nor contaminant compounds are removed. This process allows the production of two different phases which can be directly applied to land or just be prepared for perfectly controlled treatment. Furthermore, this process generates an odourless solid phase with a dry matter fraction up to 30% simplifying the handling procedure. In addition, this separation practice ensures the homogeneous solid phase can be further treated, for example, generating compost, which can be used as fertilizer, fuel or even sold over long distances with lower transport costs due to its higher density. On the other hand, the liquid phase is clarified, and is usually stored to be treated prior to discharge or reuse in cleaning or directly irrigated. Furthermore, the solid fraction reduction allows the liquid phase to be biologically treated. This helps reducing storage volume and to avoid forming manure crusting. There are several technologies to carry out this separation process, being the most common those cited below:

(a) Grid separation
(b) Screw press separation
(c) Sieving separation
(d) Press filter separation
(e) Separation by centrifugation
(f) Drum filter separation
(g) Separation by decantation.

Table 3.3 Solid–liquid separation efficiencies

Technology	Separation efficiency (%)
Sedimentation	31–56
Centrifugation	61
Sieving	44
Pressure filtration	37

The separation efficiencies of these technologies differ to a great extent from one another. Table 3.3 shows typical efficiencies values of most commonly used solid–liquid separator technologies (Hjorth et al. 2010).

Efficiencies are usually improved through physicochemical processes by adding additives that allow the coagulation and flocculation of solids for effective physical post-separation. Therefore, the choice of one technology over another depends on several other factors such as cost, particle size, pH, quantity as well as concentration of organic and inorganic components in the slurry at the feeding inlet.

Apart from that, there are also some other widely used pre-treatment practices such as the slurry acidification by adding chemical compounds. Slurry acidification consists of adding an acidic compound to the slurry to reduce the pH down to the 5–5.5 range. This increases not only the ammonium concentration by the transformation of ammoniacal nitrogen but also reduces ammonia emissions by up to 70%, as well as reducing odours (Fangueiro et al. 2015). In this case, in dairy farms, the slurry is usually mixed with zeolite, aluminium sulphate or sulphuric acid in stirring tanks which allow correct aeration in order to achieve the target pH. Previous studies have estimated that approximately 5 kg of concentrated sulphuric acid per tonne of slurry are required in this process. On the other hand, an addition of 2.5 and 6.25%wt. of zeolite reduces nearly 60 and 50% of ammonia emissions, respectively (Lefcourt and Meisinger 2001). In this way, a solid fraction in suspension and liquid fraction is pumped to storage tanks as final product. The so-obtained manure usually exhibits better fertilizing properties than the initial slurry. Today, the limitations of this technology are the operational cost due to the need for a meticulous handling by specialised workers and the need of chemical additives, which in turn results in higher costs. Finally, it must be noted that it is not totally clear whether the application of acid slurry to the soil may lead to any pollution swapping in the long term (Fangueiro et al. 2015).

After pre-treatment of the slurry, the products can either be directly applied or reused, but in most cases treatment processes must be applied to eliminate traces of nutrients and bad odours not separated in the stages discussed above. In the next paragraph, a description of each phase treatment is presented.

3.5.1.3 Solid-Phase Treatment

The most common solid-phase treatment technique is composting. This technique is based on the residual organic matter degradation to obtain a value-added product called compost. This product usually preserves well the nutrients because it is dry, odourless and pathogen-free. In addition, due to its high density, it allows for easier storage and handling, hence reducing the overall transportation costs. Additionally, due to its physicochemical properties, it can be used not only as fertilizer but also as a dairy mattress without affecting milk yield. Some studies have demonstrated that such compost reuse can be efficiently performed, being an eco-friendly technology with no extra emissions in addition to cost reduction (Zhang et al. 2019).

Composting is a solid-state fermentation where a biochemical process is carried out by micro-organisms and nematodes (Kaiser 1996). Most of the process is developed under aerobic conditions by thermophiles that hydrolyse the organic matter into humus (Onwosi et al. 2017). The treatment is developed spontaneously by the first bio-oxidative phase of aerobic degradation of organic substrates in thermophilic form (active phase). This phase is usually exothermic, leading to temperature increase, which in turn ensures drying by evaporation of water via aeration. Therefore, the mass is usually gathered into heaps to preserve the fermentation heat to guarantee higher temperatures and consequently faster reaction rates (Satyanarayana and Grajek 1999). Additionally, temperatures above 55 °C are recommended to guarantee hygienic conditions since pathogens and parasites are annihilated under those conditions. However, it should be kept in mind that the maximum temperature must be under 71 °C to prevent the elimination of thermophilic microbial population (Ravindran and Sekaran 2010). Subsequently, a maturation phase takes place where it is cooled down to ambient temperature, allowing the generation of fulvic and humic substances. Another key parameter is the moisture which must be kept between 40 and 65%wt. (Onwosi et al. 2017) to ensure an adequate metabolism. Higher contents of moisture could close mass pores turning it in a closed system, which can lead to anaerobiosis.

Alternatively, when manures are solid, carbon-rich plant structuring materials, such as straw, shavings, rubber beds and sawdust, are added in order to obtain an optimum C/N ratio. This C/N ratio is a crucial factor, where C is used as energy and N is used as cell structurer. The optimal C/N rate value must be up to 25–30 to sustain micro-organism population and evolution. As C/N ratio indicates the final compost maturity or stability, values below 25 are recommended, under 20 is suitable and under 15 is desirable. Recently, a new strategy for dairy manure composting with rice-straw has been proposed. This strategy shows a balanced nutrient with lower C/N ratio evolution of around 21.7 after 45-day fermentation, reaching 16.5 at the end of the process, improving also the pile temperature (Zhou et al. 2015).

This process is simple, although it requires large areas to produce compost piles with enough contact surface for aeration to prevent anaerobic processes that generate bad odours or stop the fermentation growth. Furthermore, the stacking space must be waterproof, flat and equipped with a leachate collection system. The generated leachate has a high organic load, ammonia-nitrogen, inorganic salts and heavy metals (Zu, Cu, etc.), which makes it environmentally troublesome (Eghball et al. 1997). Likewise, it must be frequently turned to ensure continuous aerobiosis. Moreover, there is a high probability of emissions of N_2O, CH_4, NH_3, H_2S, odours and other volatile compounds that can be washed away by aeration. This is a highly relevant topic and several research groups have been looking for solutions to reduce gas emissions by using a range of additives. For instance, medical stone, zeolite, bamboo biochar, and wood vinegar reduce the CH_4 emissions by 74% and N_2O by 69% (Mao et al., 2019).

Regarding the economic impact, composting is a rather low-cost process, with an estimated application cost of approximately between 15 and 30$ per cubic yard of produced compost, though it is a time-consuming process that usually takes between 10 and 20 weeks. To date, some new advances in composting have been reported in the search for additives to shorten the composting cycle time. Unfortunately, this strategy is not feasible yet due to additive cost-ineffectiveness. An alternative strategy focuses on biological treatments, which have been developed at lab-scale to optimise bioreactor designs to reduce time and emissions (Onwosi et al. 2017).

3.5.1.4 Treatments of the Liquid Phase

There are already several technologies for the treatment of separated liquid phase. The most commonly used are nitrification–denitrification, aerobic treatment and anaerobic ammonium oxidation (anammox) techniques. These technologies are discussed as follows.

(a) **Nitrification–Denitrification (NDN)**

NDN is a biological transformation of organic and ammoniacal nitrogen into non-polluting inert nitrogen gas. This is achieved by the combination of the oxidative and anoxic phases (autotrophic and heterotrophic nitrification). Autotrophic bacteria consume CO_2 and bicarbonates (inorganic sources) under aeration, while heterotrophic bacteria consume organic compounds anaerobically (Rajagopal and Béline 2011). This reduces odours, organic matter and removes nitrogen up to 70% which improves manure management. This technique is only applicable to slurries with total solid concentration not exceeding 6% and it should be noted that sludge is generated as by-product and needs to be managed independently. On the other hand, bad management of the global process would entail high potential environmental risk since its sensitivity to toxics and inhibitors can generate N_2O. Besides, the process is complex and depends on many parameters such as organic carbon and solid composition, NDN index, temperature, aeration system and regime type. For this reason, NDN processes must be carried out by qualified workers to guarantee an adequate management without producing NH_3 and N_2O. The investment and operation costs are very high since a continuous aeration of the medium is required, although it must be highlighted that this is the best solution for those farms which have no alternative solution to manage nitrogen excess.

(b) **Aerobic treatment**

Aerobic digestion involves the biological degradation of organic matter in the presence of oxygen. This oxygen is transmitted to the solution either via bubbling systems or by mass transfer through agitation in the free surface in bioreactors or in aerated lagoons. This technique seeks to reduce organic matter and fixed nitrogen (ammonia), where a fraction of this component passes to the solid phase. The process is developed by aerobic micro-organisms that grow forming flocs which can be physically separated very easily to obtain a stable product free of any bad smells due to pathogens reduction.

When the process is not supported by a correct nitrification–denitrification, high ammonia emissions usually are produced. On the other hand, to guarantee the process called autoheated thermophilic aerobic digestion (ATAD), which ensures high standards of hygienic quality as is highly efficient removing pathogens. To achieve this, thermophilic temperatures in the 55 and 65 °C range are preferred, and this can be carried out with the produced heat coming from the decomposition process. Consequently, this process is difficult to implement in cold areas due to large heat losses. Among the further advantages and in addition to the germicidal effect, the process is simple, and the technology requires small bioreactors as the reaction rates are rather high (Juteau 2006). However, this process is time-consuming and usually requires between 3 and 5 days, although including storage it may take up to 6 weeks to avoid the emission offensive odours. The nitrogen released to the atmosphere as ammonia may be reduced when nitrification is promoted at high aeration levels due to the prolonged treatment, though N_2O may also be produced (Burton 1992).

It should be noted that the overall cost is determined by the aerator or sparger technology, in addition to the operating costs associated to electricity consumption. The process efficiency is often expressed as kg O_2 per kWh electricity consumed, and aerators are considered efficient if they achieve performances above 1 kg O_2 per kWh. Spargers are more efficient, but their use is not widespread due to their specific aeration capacity (IAEA 2008). It is important to point out that the quality of the organic fertilizer is similar to that of compost. Furthermore, in contrast with the NDN, nitrifiers are fast growers and not too sensitive to inhibition which means that it is robust (Juteau 2006). Finally, it should be emphasised that considerable CH_4, offensive odours and reactive organic matter reduction can be achieved, though there is an increase of gaseous N losses as NH_3 plus the N_2O emissions (IAEA 2008).

(c) **Anammox**

In this case, the ammonium nitrogen is converted into inert nitrogen as in the previous case by autotrophic treatment, but with partial nitrification by oxygen injection and denitrification due to ammonium oxidation by anammox bacteria. These bacteria can oxidise ammonia under anaerobic conditions, using nitrite and producing nitrogen gas, hence it is called partial nitration. This process uses CO_2 as its C source to produce biomass ($CH_2O_{0.5}N_{0.15}$) and NO_2 as an electron acceptor for NH_4^+ oxidation but also as an electron donor for the reduction of carbon dioxide (Kuenen 2008; Szogi et al. 2015). This process produces little sludge with a very low energy cost, but given the difficulty of controlling the process, this system is one of the least common of those discussed above and it is still under research and development.

Kindaichi et al. (2016) investigated that when the digester liquid was fed to anammox reactors, the NH_4^+ and NO_2 removal efficiencies decreased by up to 32 and 42%, respectively.

3.5.2 Anaerobic Treatment

There are some other treatment technologies which consist of the slurry biological decomposition under anaerobic conditions, transforming it into a digestate and energy. This process is carried out by means of the biomethanation inside biodigesters, where a biological treatment is developed to obtain, on one hand, biogas composed of bio-methane and CO_2, and on the other hand, a bio-stabilised product called digestate which is applied as land fertilizer or food in vermiculture. As it is more stable and homogeneous, handling and further deposition in lands is facilitated. It is also well-known that this digest contains about 20% higher ammonium concentration than that not treated by anaerobic digestion (Foged et al. 2011). Also, the so-obtained biogas usually has a composition ranging from 55 to 70% methane, 36 to 45% CO_2, from <1 to 17% nitrogen and other traces of gases (Monnet 2003; Rasi et al. 2007). This biogas is usually destined to produce heat energy either by direct sale to consumers, or by generating heat to meet the farm own electricity needs for self-consumption and for sale on the grid. This anaerobic treatment is carried out by micro-organisms present in the slurry. This process takes place anaerobically in the digesters. The overall process is divided into four stages (Monnet 2003; Demirbas and Balat 2009): (1) hydrolysis, (2) acidogenesis, (3) acetogenesis and (4) methanogenesis.

1. **Hydrolysis**: At this stage the bacterial exoenzymes hydrolyse the particles and complex molecules into simpler soluble compounds, that is, the simple intermediate products obtained (proteins, carbohydrates and lipids) are transformed into soluble monomers. This is usually the limiting stage of the whole process, that is, when the production of CO_2 starts, and the pH usually varies between 5.3 and 6.7.
2. **Acidogenesis (Fermentation)**: At this point, the soluble compounds obtained in the previous hydrolysis process are fermented to produce compounds that can be consumed by methanogenic bacteria (acetic acid, formic acid, hydrogen and other reduced organic compounds). During this fermentation process, CO_2, H_2S and NH_3 are produced, giving rise to bicarbonates and organic acids, with pH between 5.1 and 6.8.
3. **Acetogenesis**: In this case, acetogenic bacteria starts producing methanogenic substrates (butyric acid, acetic acid, propionic acid and hydrogen) from the decomposition of organic acids and nitrous compounds by oxidation. The pH is usually between 6.6 and 6.8.
4. **Methanogenesis**: In this last stage, micro-organisms come into action to produce bio-methane in large volumes by digestion of proteins, amino acids, cellulose

and resistant materials. About 90% of all the generated methane in the overall process is produced in this stage. The pH varies between 6.9 and 7.4.

It should be noted that the so-produced biogas must be purified before combustion to improve its calorific value and to comply with the technical specifications of some of the most demanding thermal generators, such is the case of the application in the automotive industry. Purification consists of removing part of the generated CO_2, water vapour and other gases in lower concentrations, such as H_2S to produce a high-quality gas. To remove these gases, various techniques can be used, such as gas scrubbing, gas absorption and gas adsorption (Osorio and Torres 2009). In other applications not subjected to fuel quality requirements, only small traces of toxic gases and water vapour that can corrode the plant facilities are removed. Some available technologies destined to remove the water vapour are gas-drying cooler process by compression, active carbon adsorber and hygroscopic salts absorbers.

Currently, there are two plant models where this treatment process is developed: mesophilic and thermophilic plants. The mesophilic plants are based on operating processes running at low temperatures between 30 and 45 °C. In contrast, the thermophilic plants usually work at temperatures between 50 and 60 °C (Wang 2014). In the first model, no additional energy is required, while in the second, external energy is required to heat the biodigester and reach these thermophilic temperatures. The use of one system or another has an important impact on the overall rate and performance of the process. In addition, in thermophilic plants, a hygienic and less viscous digestate is usually obtained, which facilitates handling. Other crucial parameters to be considered when choosing the most suitable model are the type of slurry processed, the composition, nutrients and mineral concentration, pH, carbon–nitrogen ratio and the installed stirrer system. The duration of slurry retention ranges from 2 to 6 weeks.

There is a wide variety of biodigesters process technologies and designs, which can be classified generally, according to the planned feeding and discharge regime, as continuous, semi-continuous and discontinuous. In addition, there is an alternative that uses more than one stage to control the hydrolysis process regardless of the methanogenesis stage. All biodigesters are typically designed hermetically to prevent gas leaks and are always well insulated for adequate thermal control.

The choice of the process regime strongly depends on the cost, the volume of slurry handled and the available space. For instance, given that the concentration of solids is a key issue for the correct evolution of the process, it is experimentally determined that the maximum level of solids concentration for continuous operation mode is limited to 12%, while discontinuous regimes allow a feed with a solid fraction of up to 40% (Monnet 2003). In the first case, given that the feed is mostly liquid, the biofertilizer obtained is also liquid, and it is usually difficult to handle and sell, whereas the second are usually dried quickly to be sold in solid form and exhibits better fertilizing potential. Different methods used to enhance biogas production can be classified into the following groups (Fangueiro et al. 2015):

- Use of additives.
- Recycling of slurry and slurry filtrate.
- Variation in operational parameters like temperature, hydraulic retention time and particle size of the substrate.
- Use of fixed film/biofilters.

In the case of dairy manure, it is rather common that an important amount of milk from milking operations is discharged to manure digesters. This milk has an important impact on the produced biogas, causing a decrease in the CH_4 content, which may imply an increase in CO_2 production. To avoid that, the maximum milk content in the digester should be limited to 3% (Wu et al. 2011).

Finally, regarding the economic viability, Lauer et al. (2018) made an economic viability of processing dairy-cow manure for either (i) the on-farm production and use of biogas to generate electricity and heat and (ii) the upgrade of biogas to biomethane, concluding that at least 3000 cows per farm are required for an economically viable anaerobic-digestion plant operation.

3.5.3 Future Prospects

Waste management is one of the hot topics in agri-food industry as legislative/normative changes have been promoted in the last decades towards more sustainable production systems. Consequently, there is a clear trend favouring research on waste management in dairy farms. Some of the most relevant emerging technologies/strategies towards highly efficient and sustainable systems are discussed next.

(a) CO_2-RFP Strategy

This is a novel and recently reported strategy which aims at treating manure whilst reducing the environmental impact by carbon reuse from a relatively minor biogenic carbon dioxide emission source, that is, the rumen fermentation processes (RFP). This pioneering approach evaluates the viability of capturing and using RFP-CO_2 as NH_3 scrubber to produce ABC fertilizer in the dairy industry, demonstrating the potential to produce over 31.6 million of tonnes (Mt) of ABC in aqueous solution using c. 18.0 Mt of CO_2 and capturing c. 6.9 Mt of NH_3, ammonia annually (Alonso-Moreno et al. 2018). The following two chapters in this book will provide an exhaustive analysis related to this strategy (see Chaps. 4 and 5).

(b) Electrocoagulation

Several recent studies are focused on improving the use of electrocoagulation processes as an alternative to treat the dairy effluents. Electrocoagulation is an electrolytic process which entails the production of coagulants by dissolving electrically either aluminium, iron electrodes or their alloys. This emerging technique carries out a destabilisation of slurry particles by applying an electric current on the soluble anodes, which promotes the aggregation of colloidal particles in an electrochemical cell. This is due to the obtained metal ions such as Fe^{2+}, Fe^{3+} or Al^{3+} which are very efficient coagulants due to their capacity to neutralise the negative charges on the colloids. These ion metals tend to react with the hydroxyl ions produced at the cathode to generate hydroxides which in turn favour the in situ formation of flocks. Finally, these flocks can be easily separated from the liquid phase (Chen 2004; Tchamango et al. 2010).

This approach achieves higher efficiencies than the conventional solid–liquid separators, that is removal efficiencies in the 95–99% range can be typically reached (Chen 2004). Tchamango et al. (2010) have studied the treatment of the dairy effluents by electrocoagulation using aluminium electrodes where they observed a satisfactory reduction of COD, turbidity, total phosphorus and nitrogen by up to 64, 89, 81 and 100%, respectively. In their study they compared this technology with that based on the use of chemical coagulants where they found out that removal efficiencies and rates are similar, but electrocoagulation is associated to lower conductivities, hence enabling the possibility of recycling this water, though exhibiting worse COD removal efficiencies. Related to this, Melchiors et al. (2016) studied the removal rates of organic matter in dairy effluents based on the measure of COD and turbidity. They concluded that by employing aluminium electrodes COD and turbidity removal efficiencies were 97.0 and 99.6%, while using iron electrodes they reached 97.4 and 99.1%, respectively. Tchamango et al. (2010) obtained lower COD efficiencies, which may be due to the higher solubility of lactose in the dairy wastewater effluent considered.

These efficiencies are achieved in compact facilities with the possibility of implementing a fully automated process. Nevertheless, it is necessary to replace the electrodes periodically in addition to rust formation at the anode which may cause efficiency drops plus the high-power consumption that this technique entails. To illustrate this, previous studies reported that the energy consumption for treating slurry with a solid fraction of up to 1.3% was 22 kWh/m^3 (Flotats et al. 2009).

(c) **Pyrolysis**

Currently, there is an emerging treatment technology based on pyrolysis processes, which consists of the thermochemical decomposition of organic matter at high temperatures under vacuum conditions to produce fuels such as syngas, pyrolysis oils, and a solid by-product called biochar. This biochar is environmentally resistant and holds potential for carbon sequestration and soil conditioning (Samolada and Zabaniotou 2014). Slow pyrolysis is developed when the temperature ranges between 350 and 500 °C under oxygen-free conditions with long gas time residence, producing mainly char and tar. In contrast, fast pyrolysis is carried out at temperatures between 700 and 1000 °C with shorter gas residence time. In this case, most of the products are gas or liquid (pyrolysis oil) depending on the desired product and its quantity. As expected, the pyrolysis process needs external energy as is based on endothermic reactions of around 100 kJ/kg (Khiari et al. 2004). Such value increases as the waste moisture increases. For instance, a slurry with 97% of moisture requires 232.3 kJ/kg (Ro et al. 2010). Consequently, a pre-treatment dewatering process in the flushed manure is recommended to reduce the energy needs. On the other hand, the calorific value of the produced gas is between 10 and 30 MJ/m^3, while the biochar and pyrolysis oil are between 20 to 28 MJ/m^3 and 12 to 25 MJ/m^3, respectively (Khiari et al. 2004; Ro, Cantrell and Hunt, 2010; Samolada and Zabaniotou 2014; Fernandez-Lopez et al. 2015).

Pyrolysis can be considered as an innovative zero-waste method exhibiting the largest potential towards the resolution of the waste management issue, compared to other alternative methods and is characterised by lower and acceptable gas emissions. Nevertheless, given the complexity of this process and the investment cost, currently there are no industrial scale plants, therefore this can be considered as an emerging technology, still under development as the viability of this sort of large-scale plants has been proven when managing at least 20,000 tonnes/year (Samolada and Zabaniotou 2014).

(d) **Phosphorous recovery and ammonia removal technologies**

There are several emerging technologies aimed at recovering phosphorous from animal manure. One of the techniques to remove and recover N and P is the crystallisation of N and P in the form of struvite (magnesium ammonium phosphate or MgNH$_4$PO$_4$·6H$_2$O), which is a slow releasing valuable fertilizer. Depending on the composition of the wastewater, struvite precipitation can be used to remove ammonia (NH$_4^+$), phosphate (PO$_4^{3-}$) or both with efficiencies above 90% (Uludag-Demirer et al. 2005). The precipitation is activated by adding Mg^{2+} to produce Mg(OH)$_2$, MgO, and so on. The pH is the most important parameter and must be kept between 9 and 10.7, which is typically achieved by adding alkaline reagents or air stripping (Nelson et al. 2003). Shu et al. (2006) reported that this technology is technically feasible and economically beneficial with a payback period below five years for a struvite plant processing 55,000 m^3/d of waste stream. They also remarked that recovery and storage of struvite will close the phosphorus loop in the soil–crop–animal–human–soil cycle and pave way towards an ecologically sustainable future. The other existing technologies of phosphorous recovery are based on a similar process, but adding Ca^{2+} from quicklime, CaO, to produce CaNH$_4$PO$_4$·4H$_2$O or apatite, Ca$_5$(PO$_4$)$_3$OH (Li et al. 2007).

It should be noted that all the cited techniques can be combined to achieve better performances. For instance, Rico et al. 2011 have used a pre-treatment technique based on flocculation to separate liquid and solid phases, reducing the digester size. Then the liquid phase was treated anaerobically in the digester under a hydraulic retention time of 0.35 days, thus reducing the COD by 74%. Finally, a struvite precipitation treatment was applied together with the digested solid phase reaching a 95% of nutrient concentration compared to the original manure.

In the search for novel techniques devoted to recycling dairy wastewater, there are some reports focused on reverse osmosis (Vourch et al. 2008). Using this approach, the treatment of the dairy wastewaters can reach recovery performances of around 90–95%. In any case, there are many novel strategies under evaluation to prove their applicability to treat manure; some of them may seem somewhat exotic as is use of algae. Nevertheless and despite being in preliminary stages, removal efficiencies of nutrients above 80% have been reported, plus an additional renewable power generation of around 333.79–576.57 kWh/d for a dairy farm with 100 adult cattle by algal biomass combustion (Prajapati et al. 2014).

3.6   Conclusion

Dairy production systems have been analysed from both technical and management points of view, allowing an evaluation of the effect of the crucial parameters on these systems and also of their impact on climate change, as the growing concerns on environmental issues demand sustainable waste management based on novel treatment processes. Thus, different techniques were explored, explaining the available methodologies, including their potential on changing the production efficiency and on mitigating environmental pollution.

On the other hand, it has been found that intense research is ongoing for the waste management in this sector, due to the promise of largely reducing the environmental impact. For this reason, the most relevant emerging technologies and strategies considered for waste treatment and management in dairy industry have been discussed. In this context, one of these emerging strategies (the CO_2-RFP Strategy) will be fully disclosed in Chaps. 4 and 5.

References

C. Alonso-Moreno et al., The Carbon Dioxide-Rumen Fermentation Processes-strategy, a proposal to sustain environmentally friendly dairy farms. J. Clean. Prod. **204**, 735–743 (2018). https://doi.org/10.1016/j.jclepro.2018.08.295

J. Andrews, T. Davison, J. Pereira, Dairy farm layout and design: building and yard design, warm climates. Ref. Mod. Food Sci. Elsevier (2016). https://doi.org/10.1016/b978-0-08-100596-5.00705-8

D.V. Armstrong, Heat stress interaction with shade and cooling. J. Dairy Sci. **77**(7), 2044–2050 (1994). https://doi.org/10.3168/jds.S0022-0302(94)77149-6

ASABE, *ASAE D384.2 Manure production and characteristics, American Society of Agricultural and Biological Engineers.* (2005). Available at http://www.agronext.iastate.edu/immag/pubs/manure-prod-char-d384-2.pdf

S. Bittman et al., *Options for Ammonia Mitigation : Guidance From the UNECE Task Force on Reactive Nitrogen.* (Centre for Ecology & Hydrology, on behalf of Task Force on Reactive Nitrogen, of the UNECE Convention on Long Range transboundary Air Pollution, 2014)

W.P. Blount, Intensive livestock fanning. Intens. livestock Farm. 612 (1968)

A.J. Brewer, T.R. Cumby, S.J. Dimmock, Dirty water from dairy farms, II: Treatment and disposal options. Bioresour. Technol. **67**(2), 161–169 (1999). https://doi.org/10.1016/S0960-8524(98)00105-9

T.G. Brownlie, W.C. Henderson, A survey of waste management on dairy farms in South-West Scotland. Agric. Wastes **9**(4), 267–278 (1984). https://doi.org/10.1016/0141-4607(84)90085-4

C.H. Burton, A review of the strategies in the aerobic treatment of pig slurry: Purpose, theory and method. J. Agric. Eng. Res. 249–272 (1992). https://doi.org/10.1016/0021-8634(92)80086-8

C. Buxadé Carbó, *Zootecnia : bases de produccion animal*. (Mundi Prensa, 1995)

D. Chadwick et al., Manure management: Implications for greenhouse gas emissions. Anim. Feed Sci. Technol. **166–167**, 514–531 (2011). https://doi.org/10.1016/j.anifeedsci.2011.04.036

G. Chen, Electrochemical technologies in wastewater treatment. Sep. Purif. Technol. **38**(1), 11–41 (2004). https://doi.org/10.1016/j.seppur.2003.10.006

M. Cockburn et al., Lower working heights decrease contraction intensity of shoulder muscles in a herringbone 30° milking parlor. J. Dairy Sci. Elsevier **100**(6), 4914–4925 (2017). https://doi.org/10.3168/jds.2016-11629

M.F. Demirbas, M. Balat, Progress and recent trends in biogas processing. Int. J. Green Energy **6**(2), 117–142 (2009). https://doi.org/10.1080/15435070902784830

Directive, 91/676/EEC Council, 91/676/EEC council directive 91/676/EEC of 12 December 1991 concerning the protection of waters against pollution caused by nitrates from agricultural sources. 91/676/EEC Coun. Direct. 25 (1882), (1991)

B. Eghball et al., Nutrient, Carbon, and mass loss during composting of beef cattle feedlot manure. J. Environ. Qual. Am. Soc. Agron. Crop Sci. Soc. Am. Soil Sci. Soc. Am. **26**(1), 189 (1997). https://doi.org/10.2134/jeq1997.00472425002600010027x

M.I. Endres, K. Schwartzkopf-Genswein, Overview of cattle production systems. Adv. Cattle Welf. 1–26 (2018). https://doi.org/10.1016/b978-0-08-100938-3.00001-2

D. Fangueiro, M. Hjorth, F. Gioelli, Acidification of animal slurry–a review. J. Environ. Manage. 46–56 (2015). https://doi.org/10.1016/j.jenvman.2014.10.001

FAO, *World Agriculture: Towards 2030/2050*. (Rome, 2006)

M. Fernandez-Lopez et al., Life cycle assessment of swine and dairy manure: Pyrolysis and combustion processes. Bioresour. Technol. **182**, 184–192 (2015). https://doi.org/10.1016/j.biortech.2015.01.140

I. Flamenbaum et al., Cooling dairy cattle by a combination of sprinkling and forced ventilation and its implementation in the shelter system. J. Dairy Sci. **69**(12), 3140–3147 (1986). https://doi.org/10.3168/jds.S0022-0302(86)80778-0

X. Flotats et al., Manure treatment technologies: On-farm versus centralized strategies. NE Spain as case study. Bioresour. Technol. **100**(22), 5519–5526 (2009). https://doi.org/10.1016/j.biortech.2008.12.050

H.L. Foged, X. Flotats Ripoll, A. Bonmatí Blasi, Future trends on manure processing activities in Europe. Technical Report No. I concerning Manure Processing Activities in Europe to the European Commission, Directorate-General Environment, (V). (2011), p. 33. http://hdl.handle.net/2117/18943

G.M. Gustafson, Effects of daily exercise on the health of tied dairy cows. Prevent. Vet. Med. **17**(3–4), 209–223 (1993). https://doi.org/10.1016/0167-5877(93)90030-W

B.G. Hansen, Robotic milking-farmer experiences and adoption rate in Jæren, Norway. J. Rural Stud. Pergamon **41**, 109–117 (2015). https://doi.org/10.1016/J.JRURSTUD.2015.08.004

M. Hjorth et al., Solid—liquid separation of animal slurry in theory and practice. A review. Agron. Sustain. Dev. Springer Netherlands, **30**(1), 153–180 (2010). https://doi.org/10.1051/agro/2009010

J. Hyde, P. Engel, Investing in a robotic milking system: A monte carlo simulation analysis. J. Dairy Sci. **85**(9), 2207–2214 (2002). https://doi.org/10.3168/jds.S0022-0302(02)74300-2

IAEA, Guidelines for sustainable manure management in Asian livestock production systems. Brain Res. (2008). https://doi.org/10.1016/0006-8993(88)91080-3

P. Juteau, Review of the use of aerobic thermophilic bioprocesses for the treatment of swine waste. Livestock Sci. **102**(3), 187–196 (2006). https://doi.org/10.1016/j.livsci.2006.03.016

J. Kaiser, Modelling composting as a microbial ecosystem: a simulation approach. Ecol. Model. **91**(1–3), 25–37 (1996)

B. Khiari et al., Analytical study of the pyrolysis process in a wastewater treatment pilot station. Desalination **167**(1–3), 39–47 (2004). https://doi.org/10.1016/j.desal.2004.06.111

T. Kindaichi et al., Effects of organic matter in livestock manure digester liquid on microbial community structure and in situ activity of anammox granules. Chemosphere **159**, 300–307 (2016). https://doi.org/10.1016/j.chemosphere.2016.06.018

T. Kristensen, K. Søegaard, I.S. Kristensen, Management of grasslands in intensive dairy livestock farming. Livestock Prod. Sci. 61–73 (2005). https://doi.org/10.1016/j.livprodsci.2005.05.024

J.G. Kuenen, Anammox bacteria: From discovery to application. Nat. Rev. Microbiol. **6**(4), 320–326 (2008). https://doi.org/10.1038/nrmicro1857

M. Kutz (ed.), *Handbook of farm, dairy, and food machinery, handbook of farm, dairy, and food machinery* (Springer, Netherlands, Dordrecht, 2007). https://doi.org/10.1007/978-1-4020-5914-8

M. Lauer et al., Making money from waste: The economic viability of producing biogas and biomethane in the Idaho dairy industry. Appl. Energy **222**, 621–636 (2018). https://doi.org/10.1016/j.apenergy.2018.04.026

A.M. Lefcourt, J.J. Meisinger, Effect of adding alum or zeolite to dairy slurry on ammonia volatilization and chemical composition. J. Dairy Sci. **84**(8), 1814–1821 (2001). https://doi.org/10.3168/jds.S0022-0302(01)74620-6

Y. Li et al., Industrial wastewater treatment by the combination of chemical precipitation and immobilized microorganism technologies. Environ. Eng. Sci. **24**(6), 736–744 (2007). https://doi.org/10.1089/ees.2005.0026

J. Lorimor, W. Powers, A. Sutton, Manure characteristics MWPS-18, Section 1. Man. Manage. Syst. Ser. 1–24 (2008)

S.A. Luostarinen, J.A. Rintala, Anaerobic on-site treatment of black water and dairy parlour wastewater in UASB-septic tanks at low temperatures. Water Res. **39**(2–3), 436–448 (2005). https://doi.org/10.1016/j.watres.2004.10.006

H. Mao et al., Effects of four additives in pig manure composting on greenhouse gas emission reduction and bacterial community change. Bioresour. Technol. Elsevier **292**, 121896 (2019). https://doi.org/10.1016/J.BIORTECH.2019.121896

M. De Marchi, M. Penasa, M. Cassandro, Comparison between automatic and conventional milking systems for milk coagulation properties and fatty acid composition in commercial dairy herds. Ital. J. Anim. Sci. **16**(3), 363–370 (2017). https://doi.org/10.1080/1828051X.2017.1292412. Taylor & Francis

M.S. Melchiors et al., Treatment of wastewater from the dairy industry using electroflocculation and solid whey recovery. J. Environ. Manage. **182**, 574–580 (2016). https://doi.org/10.1016/j.jenvman.2016.08.022

M.R. Mondaca, Ventilation systems for adult dairy cattle. Vet. Clin. North Am. Food Anim. Prac. **35**(1), 139–156 (2019). https://doi.org/10.1016/j.cvfa.2018.10.006

F. Monnet, A final report on 'An Introduction to Anaerobic Digestion of Organic Wastes'. Carbon (2003). Available at http://www.biogasmax.co.uk/media/introanaerobicdigestion__073323000_1011_24042007.pdf

N.O. Nelson, R.L. Mikkelsen, D.L. Hesterberg, Struvite precipitation in anaerobic swine lagoon liquid: Effect of pH and Mg: P ratio and determination of rate constant. Bioresour. Technol. **89**(3), 229–236 (2003). https://doi.org/10.1016/S0960-8524(03)00076-2

C.O. Onwosi et al., Composting technology in waste stabilization: On the methods, challenges and future prospects. J. Environ. Manage. 140–157 (2017). https://doi.org/10.1016/j.jenvman.2016.12.051

F. Osorio, J.C. Torres, Biogas purification from anaerobic digestion in a wastewater treatment plant for biofuel production. Renew. Energy **34**(10), 2164–2171 (2009). https://doi.org/10.1016/j.renene.2009.02.023

S.K. Prajapati et al., Algae mediated treatment and bioenergy generation process for handling liquid and solid waste from dairy cattle farm. Bioresour. Technol. **167**, 260–268 (2014). https://doi.org/10.1016/j.biortech.2014.06.038

R. Rajagopal, F. Béline, Nitrogen removal via nitrite pathway and the related nitrous oxide emission during piggery wastewater treatment. Bioresour. Technol. **102**(5), 4042–4046 (2011). https://doi.org/10.1016/j.biortech.2010.12.017

S. Rasi, A. Veijanen, J. Rintala, Trace compounds of biogas from different biogas production plants. Energy **32**(8), 1375–1380 (2007). https://doi.org/10.1016/j.energy.2006.10.018

B. Ravindran, G. Sekaran, Bacterial composting of animal fleshing generated from tannery industries. Waste Manag **30**(12), 2622–2630 (2010). https://doi.org/10.1016/j.wasman.2010.07.013

D.J. Reinemann, M.D. Rasmussen, Milking parlors. Encycl. Dairy Sci. Elsevier 959–964 (2011). https://doi.org/10.1016/b978-0-12-374407-4.00361-7

C. Rico, H. García, J.L. Rico, Physical-anaerobic-chemical process for treatment of dairy cattle manure. Bioresour. Technol. 2143–2150 (2011). https://doi.org/10.1016/j.biortech.2010.10.068

K.S. Ro, K.B. Cantrell, P.G. Hunt, High-temperature pyrolysis of blended animal manures for producing renewable energy and value-added biochar. Ind. Eng. Chem. Res. **49**(20), 10125–10131 (2010). https://doi.org/10.1021/ie101155m

J.S. Rovira et al., Code of good agricultural practice and water pollution. Fertil. Environ. 569–572 (1996). https://doi.org/10.1007/978-94-009-1586-2_101

M.C. Samolada, A.A. Zabaniotou, Comparative assessment of municipal sewage sludge incineration, gasification and pyrolysis for a sustainable sludge-to-energy management in Greece. Waste Manag **34**(2), 411–420 (2014). https://doi.org/10.1016/j.wasman.2013.11.003

T. Satyanarayana, W. Grajek, Composting and solid state fermentation. Thermophilic Moulds Biotechnol. 265–288 (1999). https://doi.org/10.1007/978-94-015-9206-2_11

L. Shu et al., An economic evaluation of phosphorus recovery as struvite from digester supernatant. Bioresour. Technol. 2211–2216 (2006). https://doi.org/10.1016/j.biortech.2005.11.005

K.M. Svennersten-Sjaunja, G. Pettersson, Pros and cons of automatic milking in Europe1. J. Anim. Sci. **86**(suppl_13), 37–46 (2008). https://doi.org/10.2527/jas.2007-0527

A.A. Szogi, M.B. Vanotti, K.S. Ro, Methods for treatment of animal manures to reduce nutrient pollution prior to soil application. Curr. Pollut. Rep. 47–56 (2015). https://doi.org/10.1007/s40726-015-0005-1

S. Tchamango et al., Treatment of dairy effluents by electrocoagulation using aluminium electrodes. Sci. Total Environ. **408**(4), 947–952 (2010). https://doi.org/10.1016/j.scitotenv.2009.10.026

S. Uludag-Demirer, G.N. Demirer, S. Chen, Ammonia removal from anaerobically digested dairy manure by struvite precipitation. Process Biochem. **40**(12), 3667–3674 (2005). https://doi.org/10.1016/j.procbio.2005.02.028

M. Vourch et al., Treatment of dairy industry wastewater by reverse osmosis for water reuse. Desalination **219**(1–3), 190–202 (2008). https://doi.org/10.1016/j.desal.2007.05.013

J. Wang, Decentralized biogas technology of anaerobic digestion and farm ecosystem: Opportunities and challenges. Front. Energy Res. (2014). https://doi.org/10.3389/fenrg.2014.00010

X. Wu et al., Anaerobic digestion of dairy manure influenced by the waste milk from milking operations. J. Dairy Sci. **94**(8), 3778–3786 (2011). https://doi.org/10.3168/jds.2010-4129

Q. Zhang et al., Characteristics and optimization of dairy manure composting for reuse as a dairy mattress in areas with large temperature differences. J. Clean. Prod. **232**, 1053–1061 (2019). https://doi.org/10.1016/j.jclepro.2019.05.397

C. Zhou et al., A new strategy for co-composting dairy manure with rice straw: Addition of different inocula at three stages of composting. Waste Manag **40**, 38–43 (2015). https://doi.org/10.1016/j.wasman.2015.03.016

Chapter 4
Quick Fire Set of Questions About CO_2 that Need to Be Answered

Carlos Alonso-Moreno and Santiago García-Yuste

Abstract The main aim of this chapter is to update the readership on different strategies that have been designed, are in development or have been proposed for reducing CO_2 in the atmosphere. This chapter does not follow the structure of the book. The authors expect that a shift in the harmonised style of the book will facilitate reading. With the help of a quick set of fire questions, the harmlessness of the CO_2 molecule, removal strategies, capture and sequestration, and its direct transformation into other useful products are discussed in the chapter.

4.1 Is the Molecule of CO_2 Dangerous?

No, it is not. CO_2 is renewable, easily handled and stored, and essentially non-toxic. The concerns arise from its mass release, which contributes very significantly to global warming and the greenhouse effect. As early as the eighteenth century, Arrhenius linked this molecule with the global heat budget (Anderson et al. 2016). Although CO_2 represents the best example of chemical inertia, what should concern us is its molecular structure. The structure of this molecule is very active in heat transfer. CO_2 is linear and symmetrical about the central carbon atom. Its permanent bonding dipole and its three vibrational modes make it an ideal heat exchange system (Fig. 4.1) (Barrett 2005).

4.2 So, Should I Be Worried About the CO_2 Concentration in the Atmosphere?

The Intergovernmental Panel on Climate Change (IPCC) reported that CO_2 concentration must be stabilised between 350 and 440 ppm by 2050 (Bernstein et al. 2007). This recommendation means that CO_2 emissions must be reduced by 30–85%.

The main source of these exorbitant emissions is world energy demand, which produces c. 85% of the total emissions by fossil fuel combustion processes (Kharecha and Hansen 2008). The high emissions reflect the growth of anthropogenic activities

© The Author(s), under exclusive license to Springer Nature Switzerland AG 2020
S. García-Yuste, *Sustainable and Environmentally Friendly Dairy Farms*,
SpringerBriefs in Applied Sciences and Technology,
https://doi.org/10.1007/978-3-030-46060-0_4

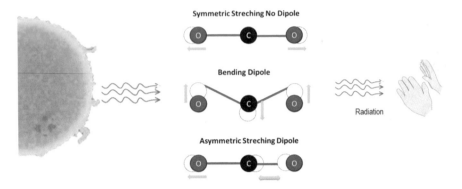

Fig. 4.1 Normal vibration modes of the CO_2 molecule

because of the explosive increase in world population (UN 2019). Accordingly, government strategies to reduce the atmospheric concentration of CO_2 focus mainly on the capture of CO_2 emissions from sources in the energy sector (UNFCC 2019). However, those efforts will not be enough (Quéré et al. 2018). As soon as possible, it is necessary to reduce any CO_2 emission, independently of its emission source. It is imperative to modify our awareness and environmental behaviour. For that purpose, you should first understand the global carbon cycle.

4.3   What Is the Carbon Cycle?

Only a deep analysis of the *global carbon budget* would lead us to understand how to control the atmospheric concentration of CO_2. We should know where the carbon reservoirs are, to take a carbon inventory and learn about the chemical transformations of carbon, independent of their source (i.e. biotic or anthropogenic) and interrelate the carbon flows.

The *global carbon budget* pinpoints the existence of well-defined carbon reservoirs, and the carbon flows between the reservoirs establish the *global carbon cycle*. Depending on the permanence of the carbon compounds, the cycle can be divided into two main categories: (1) *Fast carbon cycle*, also called '*short-term carbon cycle*', where carbon exchange between reservoirs flows easily. The carbon compounds are stored in a given carbon sink for a period lasting from days to thousands of years. (2) *Slow carbon cycle*, or '*long-term carbon cycle*', which corresponds to a carbon permanence measured in millions of years (NASA 2019).

Carbon exchange is carried out via chemical, physical, geological and/or biological transformations. Before the *Anthropocene era*, there was a balanced exchange between the carbon reservoirs. Figure 4.2 illustrates a carbon tetrahedral diagram (CTD) for a *pre-industrial* emissions *era* which might help understand the global carbon system. Each carbon reservoir is placed at the vertex of a tetrahedron, while

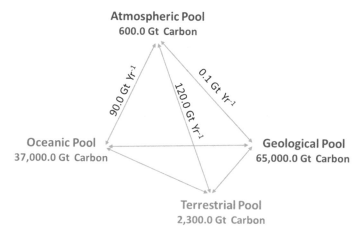

Fig. 4.2   Carbon tetrahedral diagram for the *pre-industrial* era

the flows exchanged between carbon pools are shown on the edges. The largest pools were the geological reservoir, with c. 65000.0 Gt of carbon, and the ocean reservoir with c. 37000.0 Gt of carbon (Quéré et al. 2018). At that time, the atmospheric/terrestrial and atmospheric/geological exchanges were well balanced. The atmospheric/oceanic exchange system was estimated at c. 90 Gt C yr^{-1} and it was controlled by phyto-plankton organisms, via processes of photosynthetic and respiration. Overall, in that era, carbon reservoir exchange was kept constant (Smith et al. 2016; Dowling 2018; Quéré et al. 2018).

Nowadays, worldwide CO_2 emissions from anthropogenic activities have broken the carbon equilibrium framework (Fig. 4.3). The carbon reservoir most affected is the atmospheric pool, which has increased to c. 860.0 Gt of C. Fossil fuel combustion irretrievably emits c. 9.4 Gt C yr^{-1} from the geological pool to the atmospheric.

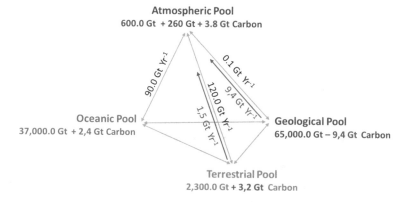

Fig. 4.3   Carbon tetrahedral diagram for the *Anthropocene* era

Additionally, agricultural and deforestation activities cause the emission of c. 1.5 Gt C from the terrestrial to the atmospheric pool (Quéré et al. 2018).

1.0 t of C is equivalent to 3.7 t of CO$_2$. Considering CO$_2$ emissions, fossil-fuel combustion and land use generate c. 38.1 Gt of CO$_2$ emissions in the atmospheric pool annually (Quéré et al. 2018). Such emissions are reallocated in c. 17.4, 8.9 and 11.8 Gt yr^{-1} of CO$_2$ in the atmospheric, oceanic and terrestrial pool, respectively. This means that the atmospheric pool is increasing its heat capacity to a worrying degree, and the ocean reservoir is suffering significant acidification, altering the marine biogeochemistry (Heinze et al. 2015). All these data are supported by the recent report from CDIAC and NOAA-ESRL, where the global carbon budget imbalance is estimated (Houghton and Nassikas 2017).

Today, we have the highest rate of CO$_2$ release to the atmospheric pool ever known, with more than 2 ppm of CO$_2$ annually, and China and India, with a very high energy demand, are still considered to be emerging economies.

4.4 Can We Do Something to Balance the Global Carbon Budget Again?

Of course, you have two options: do nothing, or do whatever you can. As explained previously, the CO$_2$ emissions from anthropogenic activities should be an important concern for all of us. With this idea in mind, let us begin by explaining the historic development and current trends in carbon dioxide capture and sequestration technologies (CCS). There are two options for CCS, using physical or chemical processes. The physical absorption is based on Henry's Law. CO$_2$ is absorbed under high pressure and low temperature and desorbed at reduced pressure and increased temperature. This technology has been widely applied to many industrial processes, including natural gas, synthesis gas and hydrogen production (Olajire 2013; Zimmermann and Kant 2017).

We have also seen considerable interest in improving CO$_2$ adsorption by chemical modification on the surface of solid materials. The basic organic group (amine) and inorganic metal oxide (alkali metal or alkali-earth metal) have both been thoroughly studied (Mondal et al. 2012). The interaction between the acidic CO$_2$ molecules and modified basic active sites on the surface facilitates CO$_2$ adsorption through the formation of covalent bonds.

The capture by solvent, membrane separation, cryogenic fractionation or adsorption using molecular sieves are the most important CCS processes reported today (Mondal et al. 2012). Spigarelli and Kawatra (2013), in an elegant review, make a detailed comparative study of all of the emerging technologies around CCS techniques: carbonate-based systems (lime-soda process), ammonia-based wet scrubbing, absorption in metal organic frameworks (MOF), zeolites, activated carbon, enzyme-based systems, and the use of ionic liquids. In a recent work, Rafiee et al. (2018) concluded that while knowledge of CSS using physical processes has reached maturity, CCS using chemical environmental strategies is still at an underdeveloped stage.

The aim of this proposal is the permanent geological storage of the captured CO_2 in deep underground porous rock by underground injection. Some of these technologies are currently available for power plants that burn fossil fuels and could play an important role in reducing CO_2 atmospheric concentration (Cuéllar-Franca and Azapagic 2015). In fact, there are now three main CCS strategies essentially working on power plants: (1) *post-combustion* in which CO_2 is captured from the fuel gases produced during combustion; (2) *pre-combustion* in which CO_2 is removed from fuel gas stream that contains mainly H_2; (3) *oxy-fuel combustion* in which fuels are burned in an environment of oxygen and recycled combustion gases (Cuéllar-Franca and Azapagic 2015). Of these, the *post-combustion* capture process is the main carbon capture technology. In terms of carbon economy amine-based compounds, and specifically monoethylamine (MEA) seems to be the best sorbent employed *in post-combustion* processes (Cuéllar-Franca and Azapagic 2015).

Even though CCS technologies could reduce over 80% of CO_2 emissions from power plants, they will not significantly help to balance the global carbon budget. Therefore, we must be ready for other environmentally sustainable options.

4.5 Are There Any Technologies Able to Capture the CO_2 Directly from the Air?

Yes, there are, and they are well studied. Mahmoudkhani and Keith (2009) proposed a green strategy called *direct air capture* (DAC) (Mahmoudkhani and Keith 2009). The proposal needs strong nucleophile reagents as an efficient CO_2-scrubber, such as amine-based or alkali-based, because CO_2 is ultra-dilute in the atmospheric pool. Even though this proposal can be developed independently from the CO_2 source, it requires very large installations. Figure 4.4 presents the DAC technologies proposed graphically and the main reaction steps (Keith et al. 2018).

The first commercial DAC plant was established in Switzerland in 2017 with a scrubber capacity of c. 1 Mt of CO_2 per year (Climeworks 2018). The most advanced commercial firm at this moment is placed in the USA, with alkali bases as scrubber. Nowadays, *Carbon Engineering* in Canada (Engineering 2018), *Climeworks* in Switzerland (Climeworks 2018), *Global Thermostat* (Global Thermostat 2019) and *Infinitree* in the USA (Infinitree 2018), *Skytree* in the Netherlands (Skytree 2019) and *Hydro Cell* in Finland (Hydrocell 2019) are companies engaged in DAC activities (Fasihi et al. 2019).

Recently, Fasihi reviewed the *state-of-the-art* of DAC technologies. It was estimated that c. 15.0 Gt of CO_2 must be captured by DAC techniques in 2050. The costs would be competitive and DAC technologies could represent a climate change mitigation solution (Fasihi et al. 2019).

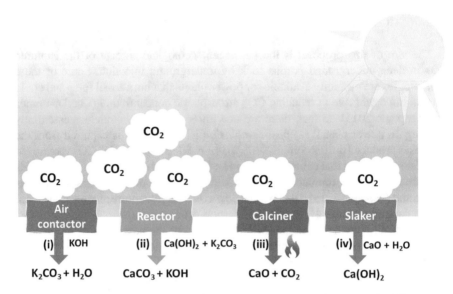

Fig. 4.4 DAC processes proposed and their main reaction steps. (i) Air contactor for CO_2 capture; (ii) Reactor for KOH recuperation; (iii) Calciner for CaO production; (iv) Slaker for $Ca(OH)_2$ synthesis

4.6   What About Making Useful Products from CO_2?

There are such technologies. This strategy is called carbon dioxide utilisation (CDU) and it is based on the 'Green Chemistry Principles' (ACS 2019). CDU techniques employ CO_2 as C1 raw material to produce chemical compounds in high demand. This proposal mainly focuses on CO_2 capture from large emissions sources. But it has important limitations in reducing CO_2 emissions (Zimmermann and Kant 2017): (1) the CO_2 sources are significantly impurified with other gaseous compounds which need prior, costly separation and purification processes (Olajire 2013), (2) the proposal requires energy (Aresta et al. 2013), (3) CO_2 activation is not a simple task and requires the presence of nucleophile chemical compounds (Appel et al. 2013), (4) most chemicals obtained in this way have relatively short lifetimes (Aresta et al. 2016), and (5) the CDU strategy must provide a lower carbon footprint than the classical synthetic methods (Zimmermann and Kant 2017).

4.7   Should We Forget About CDU Strategies?

No, we should not. Some consider that CDU is far from being an alternative to CCS based on the idea that there are not enough chemicals needed, on a comparable scale, with current CO_2 emissions (>35 Gt/year) (Omae 2012). But this is wrong!

The additional environmental and social benefits of CDU strategies must compel us to consider them as atmospheric decarbonisation tools, apart from the economic benefit (Aresta et al. 2013; Armstrong and Styring 2015; Jones et al. 2015). CDU strategies reduce the costs associated with CCS processes by transforming some of the '*waste* CO_2' into '*working carbon*' (Aresta et al. 2016). Atmospheric CO_2 should be considered a part of the *slow carbon cycle* due to its chemical inertia (Quéré et al. 2018), and any environmental strategy must be reinforced.

Daresbourg, in 2010, a reference researcher in the field of sustainable chemistry, stated that '*Clearly, processes that utilize CO_2 to produce products that meet all of the requirements of safety, performance, and cost with respect to alternative processes should be aggressively pursued*', which highlights his visionary approach to the importance of these strategies (Darensbourg 2010).

One of the best metrics for evaluating the efficiency of an environmental strategy is the number of patents published. To date, there are more than 3000 patents associated with CDU strategies (Li et al. 2013; Norhasyima and Mahlia 2018).

4.8 What Are the Most Important CDU Strategies?

Consider the CDU-related technologies patented. More than 50% of strategies patented throughout the period (2006–2010) are related to CDU, which shows the extent of academic and industrial interest in the development of such technologies (Li et al. 2013). The USA, China, Canada and Japan are the most productive countries with 60% of the total of the patents (Li et al. 2013; Norhasyima and Mahlia 2018). Of all the technologies patented, 54% is related to biomass and chemical production, 25.8% to enhance oil recovery (EOR) and coal-bed methane (ECBM), 16.3% regarding biological algae cultivation, and 3.4 and 1.5% to mineral carbonation and enhanced geothermal systems (EGS) (see Fig. 4.5) (Norhasyima and Mahlia 2018). However, the viability of each strategy must be subsequently analysed, in terms of reducing the CO_2 emissions and economic costs (Fasihi et al. 2019). To sustain any CDU implementation, a cheaper production cost than the conventional product must be achieved, the quantity of products obtained cannot exceed their market demand and the raw materials needed for the process must be abundant.

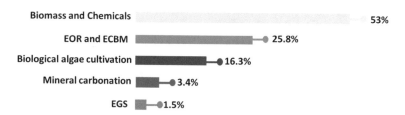

Fig. 4.5 Percentage of CDU-related technologies patented from 2008 to 2017 (Norhasyima and Mahlia 2018)

Among the most important current CDU processes, it is worth noting the world's largest commercial CDU plant for producing methanol, in Iceland, which produces c. 4000 metric tonnes per year (Recycling 2018), and the Petra Nova company in Texas (U.S.) designed to capture c. 16 Mt of CO_2 for methanol production (Zitelman et al. 2018).

4.9   Can We Use the CO_2 After Storage?

Among all the CCS technologies, the use of amine-based solvents is the most mature and low-cost technology in industry for sequestering CO_2 (Yang et al. 2017). These technologies are exclusively focused on *post-combustion* CCS processes. Among the amine-based solvents, NH_3 is the most effective CO_2 scrubber, ahead of MEA, DEA or MDA (Gonzalez-Garza et al. 2009; Olajire 2010). The advantage of NH_3 is its higher capacity to capture CO_2 and its direct transformation to $(NH_4)HCO_3$ (ABC) (Li et al. 2003). ABC has been widely used as a fertilizer in China (Li et al. 2003).

The so-called *chilled ammonia capture process* (CAP) patented by Gal (2006) for the conversion of CO_2 to $(NH_4)HCO_3$ is a representative example of the advantages of CDU technologies based on NH_3 as scrubber. CAP is low cost and accessible worldwide, the final product is chemically stable, its CO_2 loading capacity is very high and CO_2 regeneration is possible at high pressures. CAP was commercialised by Alstom Power, in 2013 (Lombardo et al. 2014).

4.10   Can You Explain a Little Bit More About the CO_2-NH_3 System?

The ternary CO_2-NH_3-H_2O system is very well known ((Versteeg et al. 1996; Li et al. 2003; Gal 2006), but its experimental optimisation is still challenging (Sutter and Mazzotti 2017; Milella et al. 2018). Scheme 4.1 shows the chemical speciation of the system. The amount of CO_2 dissolved in the aqueous solution depends on the Henry's Law constant, which is related to pH, temperature and ionic strength. High temperature increases the constant and high pH raises the effectiveness of the CO_2 absorption.

To date, there are three mechanisms proposed to explain the CO_2-NH_3-H_2O system: (1) carbamate formation mechanism, (2) zwitterion mechanism, and (3) termolecular mechanism (Scheme 4.2). The first mechanism was reported by Caplow and Danckwerts some decades ago (Caplow 1968; Danckwerts 1979). The formation of carbamate species and their subsequent hydrolysis justify the generation of ABC. Versteeg suggested the zwitterion mechanism in 1996 (Versteeg et al. 1996; Ramachandran et al. 2006). This mechanism corresponds to a Caplow mechanism simplification where the base reagent is H_2O. Finally, the termolecular mechanism

Aqueous ammonia system

$$NH_{3(g)} + H_2O_{(l)} \rightleftharpoons NH_3 \cdot H_2O_{(aq)} \rightleftharpoons NH_{3(aq)} \qquad (1)$$

$$NH_{3(aq)} + H_2O_{(l)} \rightleftharpoons NH_4^+{}_{(aq)} + OH^-{}_{(aq)} \qquad (2)$$

Aqueous CO_2 system

$$CO_{2(g)} + H_2O_{(l)} \rightleftharpoons CO_2 \cdot H_2O_{(aq)} \rightleftharpoons CO_{2(aq)} \qquad (3)$$

$$CO_{2(aq)} + H_2O_{(l)} \rightleftharpoons HCO_3^-{}_{(aq)} + H^+{}_{(aq)} \qquad (4)$$

$$HCO_3^-{}_{(aq)} \rightleftharpoons CO_3^{-2}{}_{(aq)} + H^+{}_{(aq)} \qquad (5)$$

Final Chemical Compounds

$$NH_4^+{}_{(aq)} + HCO_3^-{}_{(aq)} \rightleftharpoons (NH_4)HCO_{3(aq)} \qquad (6)$$

$$2\,NH_4^+{}_{(aq)} + CO_3^{-2}{}_{(aq)} \rightleftharpoons (NH_4)_2CO_{3(aq)} \qquad (7)$$

Scheme 4.1 Chemical speciation of the CO_2-NH_3-H_2O system. Equations (1 and 2) describe the aqueous ammonia system and Eqs. (3–5) the CO_2 aqueous system. Finally, Eqs. (6 and 7) show the chemical compounds yielded by the ternary CO_2-NH_3-H_2O system

consists of a single-step formation of the carbamate, after simultaneous reaction of ammonia with CO_2 and H_2O (Mao et al. 2019).

As early as 2011, Herzog explained the potential benefits of large giga-tone CO_2-NH_3-H_2O system for CO_2 sequestration (Herzog 2011), but today, only 17 industrial scale projects are technically viable. In 2018, the sequestration of a mere c. 30.0 Mt of CO_2 by CO_2-NH_3-H_2O system has been recorded (Bui et al. 2018).

4.11 Are There Any CDU Strategies Proposed for Minor Sources?

Data reveal that more than half of all CO_2 emissions originate from small and dispersed sources, such as agricultural operations or the transportation sector (Knapp et al. 2014). However, there are few proposals for addressing the problem and none are in industrial development. As an example of potential strategies planned for minor sources, consider the 'CO_2-AFP Strategy' as a representative example (Alonso-Moreno and García-Yuste 2016). Total world CO_2 emissions generated by alcohol

Carbamate Formation Mechanism

$$2\,NH_3 \;+\; CO_2 \;\rightleftharpoons\; NH_2COO(NH_4) \qquad (10)$$

$$NH_2COO(NH_4) \;+\; H_2O \;\rightleftharpoons\; (NH_4)HCO_3 \;+\; NH_3 \qquad (11)$$

$$NH_3 \;+\; H_2O \;\rightleftharpoons\; NH_4OH \qquad (12)$$

$$(NH_4)HCO_3 \;+\; NH_4OH \;\rightleftharpoons\; (NH_4)_2CO_3 \;+\; H_2O \qquad (13)$$

$$(NH_4)_2CO_3 \;+\; H_2O \;+\; CO_2 \;\rightleftharpoons\; 2\,(NH_4)HCO_3 \qquad (14)$$

Zwitterion Mechanism

$$NH_3 \;+\; CO_2 \;\rightleftharpoons\; NH_3{}^{+}COO^{-} \qquad (15)$$

$$NH_3{}^{+}COO^{-} \;+\; NH_3 \;\rightleftharpoons\; NH_2COOH^{-} \;+\; NH_3 \qquad (16)$$

$$NH_2COO^{-} \;+\; NH_4{}^{+} \;+\; 2\,H_2O \;\rightleftharpoons\; (NH_4)HCO_3 \;+\; NH_3 \;+\; H_2O \qquad (17)$$

Termolecular Mechanism

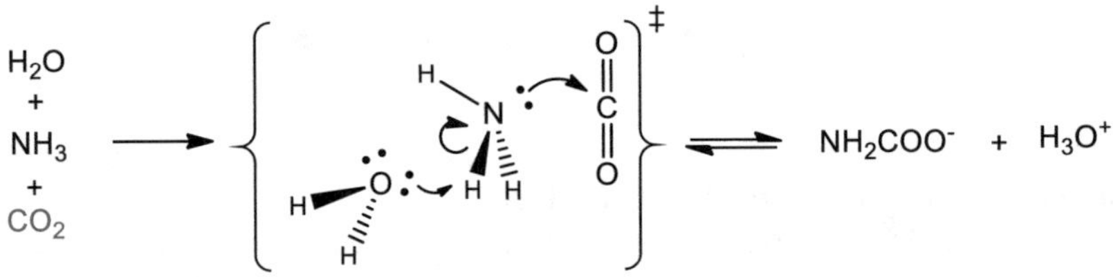

Scheme 4.2 Chemical equations to define carbamate formation mechanism, zwitterion mechanism and termolecular mechanism for ABC generation

fermentation processes (AFP) are c. 12.7 Mt in wine and spirit production (Alonso-Moreno and García-Yuste 2016). The 'CO_2-AFP Strategy' consists of using NaOH as a strong nucleophilic reagent to activate the CO_2 released during the fermentation processes. One mole of soda ash (Na_2CO_3) in aqueous solution would be obtained for each mole of CO_2 released by fermentation, and 2 mol of NaOH used as reagent (see illustration of the strategy in Fig. 4.6). In 2013, the total world soda ash (Na_2CO_3) production was c. 51.3 Mt, distributed into 25% from the mineral trona, and 75% from synthetic Solvay ammonia soda processes. The 'CO_2-AFP Strategy' is a good example of an atomic economy process. From a thermodynamic point of view, the conversion of CO_2 into soda ash is highly favoured. The implementation of this CDU

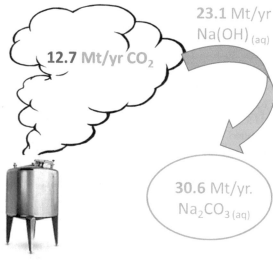

Alcohol Fermentation Process (AFP)

Fig. 4.6 Environmental potential of the use of CO_2 from alcoholic fermentation processes: The 'CO_2-AFP Strategy' (Alonso-Moreno and García-Yuste 2016)

strategy could give rise to the production of 30.6 Mt of soda ash per year, more than 50% of the annual production worldwide, and avoid c. 30.6 Mt of CO_2 emissions.

The 'CO_2-AFP Strategy' fits in perfectly with the concept of *Green Motion*™ (Phan et al. 2015) since the CO_2 would be used as a non-toxic and non-hazardous raw material to generate, under reasonable conditions, the desired product, and water as a by-product in a 100% atomic economy. The advantages of this strategy are: (1) high purity of CO_2 (c. 100%), (2) mild condition reactions (i.e. room temperature, normal pressure), (3) little equipment required, (4) high effectiveness in CO_2 transformation, (5) generation of one of the top 10 industrial commodities (Na_2CO_3), (6) sodium compounds are compatible and environmentally friendly in the food industry and broadly employed in typical general processes (i.e. cleaner or neutralisation baths), (7) no other by-product will be produced apart from H_2O, and (8) the energetic balance shows an exothermic process (Alonso-Moreno and García-Yuste 2016).

4.12 What Is the Meaning of NETs?

A few years ago, the Royal Society of Chemistry (RSC) founded the *CO_2Chem network*. The main aim of this organization was to work on these three main areas for preventing CO_2 emissions: (1) CO_2 transformation to Bulk Chemicals and fuels, (2) Carbon Capture, (3) CO_2 transformation to Fine Chemicals. Imperial College

London later reported a group of available technologies, the so-called '*Negative Emissions Technologies*' for the removal of CO_2 from the atmosphere (some important CDU strategies are included among them) (Alonso-Moreno and García-Yuste 2016). However, the concept of NETs goes beyond the development of technologies. Perhaps, the statement '*Intentional human efforts to remove CO_2 emissions from the atmosphere employing these technologies*' used by Minx in 2018 to define NETs could help to understand the meaning of this movement (Minx et al. 2018). Minx also proposed a classification for NETs: (1) Technologies that use the photosynthetic process to capture CO_2 from the atmosphere, (2) technologies that transform the CO_2 in the atmosphere (Rafiee et al. 2018; Minx et al. 2018).

Regarding the first point, afforestation and reforestation (AF), bioenergy with carbon capture and storage (BECCS), sugarcane production and enhanced biological fixation and CO_2 capture using fast-growing biomass by biological methods (OF) are the most important. AF absorbs CO_2 through plant growth (Smith et al. 2016). Although AF requires low-cost technologies, large areas are needed to absorb significant quantities of CO_2. It is estimated that the global potential of AF capacity is c. 1.1–3.3 Gt C yr^{-1}. But this CO_2 capture would require 320 to 970 MHa of land, which is equivalent to c. 20–60% of the arable land. Smith, in 2015, noted another important limitation, the additional fertilization processes required. This would increase the N_2O emission rate, which hampers the beneficial GHG effects on CO_2 reduction (Smith et al. 2016).

BECCS encourages the use of bioenergy with carbon capture and storage at the expense of fossil fuel combustion (Smith et al. 2016). This strategy works on the assumption that trees absorb carbon, and if you burn them for energy, and then capture and bury the emissions below ground, you remove CO_2 from the atmosphere. This NET is easy to implement because growing biomass is a tangible technology and CCSs are mature projects nowadays. However, there are reasons that BECCS is not a viable solution today: (1) It would need very large expanses of land, (2) The dedicated crops grown for BECCS would compete with land to grow food, (3) BECCS would reduce biodiversity if implemented at scale (Smith et al. 2016).

Although photosynthesis is not a very efficient process, every day plants transform the sun's energy into organic chemical compounds (Goeppert et al. 2012). Expanding sugarcane production in the world for conversion to ethanol could current CO_2 emissions (Goeppert et al. 2012). However, since the average efficiency is limited to 0.5–2.0% in most crops, this strategy would involve the conversion of hundreds of thousands of square miles for sugarcane production (Rafiee et al. 2018).

Chiang and Pan (2017) justified the inclusion of microalgae plant systems as part of NET technologies (Chiang and Pan 2017). Their efficiency in transforming CO_2 into biochemicals and biofuels is higher than alternative plant systems based on energetic crops. The quantity of CO_2 captured by microalgae is c. 150 Mt Ha^{-1} yr^{-1} by photobioreactors and 50–70 Mt Ha^{-1} yr^{-1} in open ponds, whereas only 3–13 Mt Ha^{-1} yr^{-1} are reported for terrestrial plant systems (i.e. grass, corn or soybeans) (Chiang and Pan 2017).

Two technologies concerning the transformation of CO_2 in the atmosphere should be highlighted. The first is DAC, which was discussed previously. The other is so-called enhanced weathering (EW) (Smith et al. 2016). This strategy proposes the use of silicate or carbonate minerals for CO_2 sequestration. Olivine or basalt are good minerals for EW in solids and oceans. However, the same problem arises in developing this strategy; that EW would require large amounts of minerals. There is an estimated need for c. 1.0–3.0 Gt of minerals per Gt of C removed. This is before considering the mining and transporting processes required before using those minerals (Smith et al. 2016).

In 2017, the RSC and the Royal Academy of Engineering published a document looking at the scientific and engineering perspectives on *greenhouse gas removal* (GGR) processes (Dowling 2018). The GGR proposal goes from increased biological uptake (i.e. AF, soil carbon sequestration, biochar, BECCS, OF), natural inorganic reactions (i.e. ocean alkalinity, enhanced terrestrial weathering, mineral carbonation) and engineered removal (i.e. DACCS, low carbon concrete) in terrestrial, oceanic or geological storage sinks. This document brings together the economic and environmental difficulties involved in the implementation of large-scale GGRs.

Haszeldine reported, in 2018, an analysis of the NETs strategies for achieving the Paris Agreement commitments (Haszeldine et al. 2018). It will require the capture of 6000 Mt CO_2 yr^{-1} to meet the 2 °C target. It is a long way from the current NETs projection to 2050, which is estimated as c. 700 Mt CO_2 yr^{-1}. NETs are theoretically worthwhile, but in practice only AF is ready for full deployment.

Figure 4.7 illustrates the hypothetical influences of NETs on the CTD (Smith et al. 2016; Quéré et al. 2018). Carbon would flow between carbon reservoirs in several ways: (a) in BECCS technologies, carbon is captured from the atmospheric pool to grow bio-energy feedstocks (terrestrial pool), (b) EW would accelerate natural weathering, removing CO_2 emissions from the atmosphere by yielding carbon species stored in the geological pool, (c) AF and sugarcane production would capture CO_2 from the atmosphere by fixing atmospheric carbon into biomass (terrestrial

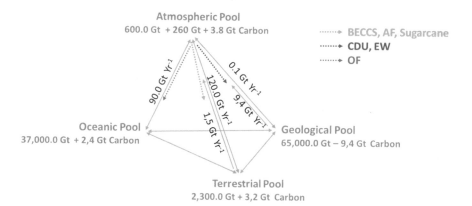

Fig. 4.7 Influence of NETs development on the CTD (Smith et al. 2016; Quéré et al. 2018)

pool), (d) OF would remove CO_2 from the atmosphere by enhancing ocean alkalinity (oceanic pool), (e) CDU would capture CO_2 from the atmospheric pool by saving carbon in the geologic pool (fossil-fuels equivalents) or by its transformation into chemicals.

4.13   How Can We Analyse the Sustainability of the Different Strategies Proposed?

For this purpose, there are two concepts that need to be understood first, *sustainability* and *circular economy*. Consider the definition of the word sustainability, given by Morelli in 2011 (Morelli 2011) from an environmental point of view: *'A condition of balance, resilience, and interconnectedness that allows human society to satisfy its needs while neither exceeding the capacity of its supporting ecosystems to continue to regenerate the services necessary to meet those needs nor by our actions diminishing biological diversity.'*

The concept of *circular economy* (CE) comes from important economics schools (i.e. Industrial Ecology, Cradle to Cradle, the Performance Economy, The Blue Economy and Biomimicry) (Wautelet 2018). This is the opposite of the concept of lineal economy. The main difference between the two concepts is their responsibility to society and environment. The linear economy only values the final product, independent of its implications. However, CE is based on the principles of designing out waste and pollution, keeping products and materials in use and regenerating natural systems (see Fig. 4.8 to visualise the differences) (Wautelet 2018).

Any new environmental proposal for removing CO_2 needs to be analysed in terms of economy and sustainability. These strategies are complex environmental systems which would require the analysis of the five key parameters and their interconnection (see circular economy diagram in Fig. 4.9). This can only be achieved from a multi-directional perspective. Could we use a symmetrical polyhedron? The tetrahedron is the simplest polyhedron, formed by four triangular faces (equilateral triangles), six straight edges and four vertices. The centroid of a solid tetrahedron is the centre of gravity. Due to the existence of five key parameters, the centroid of the polyhedron will be used to generate the basis of any analysis. As was previously mentioned, there are five key parameters: raw material, production, consumption, waste management and waste prevention. Any sustainable production process must use proper raw materials and yield products demanded by society. The waste prevention and waste management must follow the sustainable requirements to preserve the environmental. Figure 4.9 shows how these key parameters are in the centroid of regular tetrahedra and how they are interconnected.

Thus, a new environmental proposal must analyse the sustainability of the raw materials (Fig. 4.9a) and the production processes (Fig. 4.9b). Furthermore, any development tackling one of these issues must factor in all the others in order to be sustainable. When studying the sustainability of raw materials (Fig. 4.9a), these

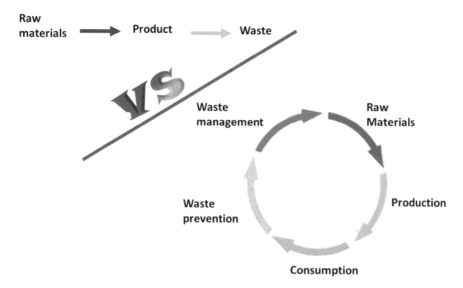

Fig. 4.8 Linear economy versus circular economy

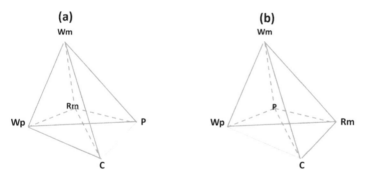

Rm = Raw materia; P = Production; C = Consumption; Wp = Waste prevention; Wm = Waste managment

Fig. 4.9 Tetrahedral economy diagrams. **a** Sustainability of raw materials. **b** Sustainability of production processes

must be employed in sustainable production, with negligible effect on the consumer market, with a strong focus on waste prevention and the best management practises.

Likewise, when the parameter analysed is a production process (Fig. 4.9b), the raw material used must be obtained by the best available techniques, generating a product that the market needs (consumption), where waste prevention must be considered in order to increase production efficiency, and finally proper waste management should guarantee the lowest environmental effect.

References

C. Alonso-Moreno, S. García-Yuste, Environmental potential of the use of CO$_2$ from alcoholic fermentation processes. The CO$_2$-AFP strategy. Sci. Total Environ. **568**, 319–326 (2016). https://doi.org/10.1016/j.scitotenv.2016.05.220

T.R. Anderson, E. Hawkins, P.D. Jones, CO$_2$, the greenhouse effect and global warming: from the pioneering work of Arrhenius and callendar to today's earth system models. Endeavour **40**, 178–187 (2016). https://doi.org/10.1016/j.endeavour.2016.07.002

A.M. Appel, J.E. Bercaw, A.B. Bocarsly et al., Frontiers, opportunities, and challenges in biochemical and chemical catalysis of CO$_2$ fixation. Chem. Rev. **113**, 6621–6658 (2013)

M. Aresta, A. Dibenedetto, A. Angelini, The changing paradigm in CO$_2$ utilization. J. CO$_2$ Util. **3–4**, 65–73 (2013)

M. Aresta, A. Dibenedetto, E. Quaranta, State of the art and perspectives in catalytic processes for CO$_2$ conversion into chemicals and fuels: The distinctive contribution of chemical catalysis and biotechnology. J. Catal. **343**, 2–45 (2016). https://doi.org/10.1016/j.jcat.2016.04.003

K. Armstrong, P. Styring, Assessing the potential of utilization and storage strategies for post-combustion CO$_2$ emissions reduction. Front. Energy Res. **3** (2015). https://doi.org/10.3389/fenrg.2015.00008

J. Barrett, Greenhouse molecules, their spectra and function in the atmosphere. Energy Environ. **16**, 1037–1045 (2005). https://doi.org/10.1260/095830505775221542

L. Bernstein, P. Bosch, O. Canziani et al., Climate change 2007 : Summary for policymakers. Hemisphere 12–17 (2007)

M. Bui, C.S. Adjiman, A. Bardow et al., Carbon capture and storage (CCS): The way forward. Energy Environ. Sci. **11**, 1062–1176 (2018)

M. Caplow, Kinetics of Carbamate Formation and Breakdown. J. Am. Chem. Soc. **90**, 6795–6803 (1968). https://doi.org/10.1021/ja01026a041

P.C. Chiang, S.Y. Pan, Carbon dioxide mineralization and utilization. (2017)

Climeworks, *About| Climeworks—Capturing CO$_2$ From Air*. (2018). http://www.climeworks.com/about/. Accessed 25 Nov 2019

R.M. Cuéllar-Franca, A. Azapagic, Carbon capture, storage and utilisation technologies: A critical analysis and comparison of their life cycle environmental impacts. J. CO$_2$ Util. **9**, 82–102 (2015)

P.V. Danckwerts, The reaction of CO$_2$ with ethanolamines. Chem. Eng. Sci. **34**, 443–446 (1979). https://doi.org/10.1016/0009-2509(79)85087-3

D.J. Darensbourg, Chemistry of carbon dioxide relevant to its utilization: A personal perspective. Inorg. Chem. **49**, 10765–10780 (2010). https://doi.org/10.1021/ic101800d

D.A. Dowling, Greenhouse gas removal. (2018)

C. Engineering, *About Us—Carbon Engineering*. http://carbonengineering.com/about-us/. Accessed 25 Nov 2019

M. Fasihi, O. Efimova, C. Breyer, Techno-economic assessment of CO$_2$ direct air capture plants. J. Clean. Prod. **224**, 957–980 (2019). https://doi.org/10.1016/j.jclepro.2019.03.086

E. Gal, *US20080072762A1—Ultra Cleaning of Combustion Gas Including the Removal of CO$_2$—Google Patents*. (2006). https://patents.google.com/patent/US20080072762A1/en. Accessed 25 Nov 2019

Global Thermostat, *About Global Thermostat—Global Thermostat*. (2019). https://globalthermostat.com/about-global-thermostat/. Accessed 25 Nov 2019

A. Goeppert, M. Czaun, G.K. Surya Prakash, G.A. Olah, Air as the renewable carbon source of the future: An overview of CO$_2$ capture from the atmosphere. Energy Environ. Sci. **5**, 7833–7853 (2012)

D. Gonzalez-Garza, R. Rivera-Tinoco, C. Bouallou, Comparison of ammonia, monoethanolamine, diethanolamine and methyldiethanolamine solvents to reduce CO$_2$ greenhouse gas emissions. Chem. Eng. Trans. 279–284 (2009)

R.S. Haszeldine, S. Flude, G. Johnson, V. Scott, Negative emissions technologies and carbon capture and storage to achieve the Paris agreement commitments. Philos. Trans. R. Soc. A Math. Phys. Eng. Sci. **376**, 20160447 (2018). https://doi.org/10.1098/rsta.2016.0447

C. Heinze, S. Meyer, N. Goris et al., The ocean carbon sink—impacts, vulnerabilities and challenges. Earth Syst. Dyn. **6**, 327–358 (2015)

H.J. Herzog, Scaling up carbon dioxide capture and storage: From megatons to gigatons. Energy Econ. **33**, 597–604 (2011). https://doi.org/10.1016/j.eneco.2010.11.004

R.A. Houghton, A.A. Nassikas, Global and regional fluxes of carbon from land use and land cover change 1850–2015. Global Biogeochem. Cycles **31**, 456–472 (2017). https://doi.org/10.1002/2016GB005546

Hydrocell, *Direct Air Capture (DAC) Appliances—Hydrocell Oy*. (2019). https://hydrocell.fi/en/air-cleaners-carbon-dioxide-filters-and-dac-appliances/dac-appliances/. Accessed 25 Nov 2019

Infinitree, *Technology—Infinitree LLC*. (2018). http://www.infinitreellc.com/technology/. Accessed 25 Nov 2019

C.R. Jones, D. Kaklamanou, W.M. Stuttard et al., Investigating public perceptions of carbon dioxide utilisation (CDU) technology: A mixed methods study. Faraday Discuss. **183**, 327–347 (2015). https://doi.org/10.1039/c5fd00063g

D.W. Keith, G. Holmes, D. St. Angelo, K. Heidel, A process for capturing CO_2 from the atmosphere. Joule **2**, 1573–1594 (2018). https://doi.org/10.1016/j.joule.2018.05.006

P.A. Kharecha, J.E. Hansen, Implications of 'peak oil' for atmospheric CO_2 and climate. Global Biogeochem. Cycles **22** (2008). https://doi.org/10.1029/2007GB003142

J.R. Knapp, G.L. Laur, P.A. Vadas et al., Invited review: Enteric methane in dairy cattle production: Quantifying the opportunities and impact of reducing emissions. J. Dairy Sci. **97**, 3231–3261 (2014)

B. Li, Y. Duan, D. Luebke, B. Morreale, Advances in CO_2 capture technology: A patent review. Appl. Energy **102**, 1439–1447 (2013)

X. Li, E. Hagaman, C. Tsouris, J.W. Lee, Removal of carbon dioxide from flue gas by ammonia carbonation in the gas phase. Energy Fuels **17**, 69–74 (2003). https://doi.org/10.1021/ef020120n

G. Lombardo, R. Agarwal, J. Askander, Chilled ammonia process at technology center Mongstad-first results. Energy Procedia 31–39 (2014)

M. Mahmoudkhani, D.W. Keith, Low-energy sodium hydroxide recovery for CO_2 capture from atmospheric air-Thermodynamic analysis. Int. J. Greenh. Gas Control **3**, 376–384 (2009). https://doi.org/10.1016/j.ijggc.2009.02.003

H. Mao, H. Zhang, Q. Fu et al., Effects of four additives in pig manure composting on greenhouse gas emission reduction and bacterial community change. Bioresour. Technol. **292**, 121896 (2019). https://doi.org/10.1016/J.BIORTECH.2019.121896

F. Milella, M. Gazzani, D. Sutter, M. Mazzotti, Process synthesis, modeling and optimization of continuous cooling crystallization with heat integration—application to the chilled ammonia CO_2 Capture Process. Ind. Eng. Chem. Res. **57**, 11712–11727 (2018). https://doi.org/10.1021/acs.iecr.8b01993

J.C. Minx, W.F. Lamb, M.W. Callaghan et al., Negative emissions—Part 1: Research landscape and synthesis. Environ. Res. Lett. **13**, 63001 (2018). https://doi.org/10.1088/1748-9326/aabf9b

M.K. Mondal, H.K. Balsora, P. Varshney, Progress and trends in CO_2 capture/separation technologies: A review. Energy **46**, 431–441 (2012). https://doi.org/10.1016/j.energy.2012.08.006

J. Morelli, Environmental sustainability: A definition for environmental professionals. J. Environ. Sustain. **1**, 1–10 (2011). https://doi.org/10.14448/jes.01.0002

NASA, *The Carbon Cycle*. (2019). https://earthobservatory.nasa.gov/features/CarbonCycle. Accessed 25 Nov 2019

R.S. Norhasyima, T.M.I. Mahlia, Advances in CO_2 utilization technology: A patent landscape review. J. CO_2 Util. **26**, 323–335 (2018). https://doi.org/10.1016/j.jcou.2018.05.022

A.A. Olajire, CO_2 capture by aqueous ammonia process in the clean development mechanism for Nigerian oil industry. Front. Chem. Sci. Eng. **7**, 366–380 (2013)

A.A. Olajire, CO_2 capture and separation technologies for end-of-pipe applications—A review. Energy **35**, 2610–2628 (2010). https://doi.org/10.1016/j.energy.2010.02.030

I. Omae, Recent developments in carbon dioxide utilization for the production of organic chemicals. Coord. Chem. Rev. **256**, 1384–1405 (2012)

T.V.T. Phan, C. Gallardo, J. Mane, GREEN MOTION: A new and easy to use green chemistry metric from laboratories to industry. Green Chem. **17**, 2846–2852 (2015). https://doi.org/10.1039/c4gc02169j

C. Quéré, R. Andrew, P. Friedlingstein et al., Global carbon budget 2018. Earth Syst Sci Data **10**, 2141–2194 (2018). https://doi.org/10.5194/essd-10-2141-2018

A. Rafiee, K. Rajab Khalilpour, D. Milani, M. Panahi, Trends in CO_2 conversion and utilization: A review from process systems perspective. J. Environ. Chem. Eng. **6**, 5771–5794 (2018)

N. Ramachandran, A. Aboudheir, R. Idem, P. Tontiwachwuthikul, Kinetics of the absorption of CO_2 into mixed aqueous loaded solutions of monoethanolamine and methyldiethanolamine. Ind. Eng. Chem. Res. 2608–2616 (2006)

C. Recycling, *CRI—Carbon Recycling International*. (2018). https://www.carbonrecycling.is/. Accessed 25 Nov 2019

Skytree (2019) Direct Air Capture - Skytree. https://www.skytree.eu/direct-air-capture/. Accessed 25 Nov 2019

P. Smith, S.J. Davis, F. Creutzig et al., Biophysical and economic limits to negative CO_2 emissions. Nat. Clim. Chang. **6**, 42–50 (2016)

B.P. Spigarelli, S.K. Kawatra, Opportunities and challenges in carbon dioxide capture. J. CO2 Util. **1**, 69–87 (2013)

D. Sutter, M. Mazzotti, Solubility and growth kinetics of ammonium bicarbonate in aqueous solution. Cryst. Growth Des. **17**, 3048–3054 (2017). https://doi.org/10.1021/acs.cgd.6b01751

UN, *World Population Prospects—Population Division—United Nations*. (2019). https://population.un.org/wpp/. Accessed 25 Nov 2019

UNFCC, *Paris Agreement—Status of Ratification*. (UNFCCC, 2019). https://unfccc.int/process/the-paris-agreement/status-of-ratification. Accessed 25 Nov 2019

G.F. Versteeg, L.A.J. Van Dijck, W.P.M. Van Swaaij, On the kinetics between CO_2 and alkanolamines both in aqueous and non-aqueous solutions. An overview. Chem. Eng. Commun. **144**, 133–158 (1996)

T. Wautelet, The concept of circular economy—its origins and its evolution (2018). https://doi.org/10.13140/RG.2.2.17021.87523

X. Yang, R.J. Rees, W. Conway et al., Computational modeling and simulation of CO_2 capture by aqueous amines. Chem. Rev. **117**, 9524–9593 (2017)

A. Zimmermann, M. Kant, CO_2 utilisation today. (2017)

K. Zitelman, J. Ekmann, J. Huston, P. Indrakanti, Carbon capture, utilization, and storage: Technology and policy status and opportunities national association of regulatory utility commissioners. (2018)

Chapter 5
The 'CO$_2$-RFP Strategy'

Carlos Alonso-Moreno and Santiago García-Yuste

Abstract This chapter of the book presents the 'CO$_2$-RFP Strategy' as an innovative proposal for carbon dioxide use in the husbandry industry. Firstly, an introduction is provided to the important concepts the reader will need to learn to navigate a 'minimal path' through the strategy. Secondly, estimates of CO$_2$ and NH$_3$ emissions from husbandry (livestock) production are derived to help understand the implications and the guiding principles of the strategy. Finally, the strategy is theoretically explained and discussed as a business model.

5.1 Introduction

CO$_2$ is renewable, easily handled and stored, and essentially non-toxic. However, the massive emission of this compound, mainly from energy generation, industry and deforestation, contributes critically to warming the atmosphere. International institutions such as the United Nations (2015), FAO (2012), IPCC (2007), NASA (2017), USEPA (2017) or IEA (2017) are becoming more responsive to sustainability concerns (Alonso-Moreno and García-Yuste 2016). A plethora of scientific publications regarding proposals to mitigate CO$_2$ emissions shows this trend in recent years (Omae 2012; Zhao et al. 2012; Reis et al. 2015; Kleij et al. 2017). As previously discussed in Chap. 4, there are two alternatives, carbon dioxide utilization (CDU) and carbon capture and storage (CCS). Recently, Amstrong and Styring reported a significantly higher estimate of CO$_2$ capture by the use of CDU as compared to the capture predicted by the CCS alternative (Armstrong and Styring 2015).

Unexpectedly, the data reveal that more than half of all CO$_2$ emissions originate from small and dispersed sources, such as agricultural operations or the transportation sector (Alonso-Moreno et al. 2018). Figure 5.1 shows the growing interest in the scientific literature concerning the environmental impact of everyday activities. Even those CO$_2$ emissions coming from a biogenic source, which have never been considered within the CO$_2$-GHG-content emissions, are being carefully analysed (Alonso-Moreno et al. 2018). Thus, novel CDU strategies to mitigate CO$_2$ emissions form relatively minor sources are very welcome.

© The Author(s), under exclusive license to Springer Nature Switzerland AG 2020
S. García-Yuste, *Sustainable and Environmentally Friendly Dairy Farms*,
SpringerBriefs in Applied Sciences and Technology,
https://doi.org/10.1007/978-3-030-46060-0_5

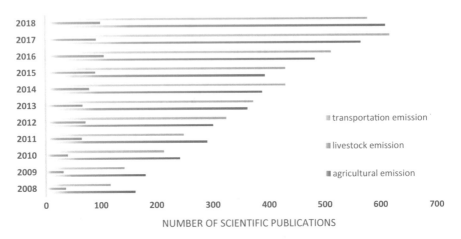

Fig. 5.1 Scientific production concerning the environmental impact of everyday activities

Ruminants generate the largest quantity of anthropogenic GHG emissions.
Emissions from husbandry processes are the dispersed source of greatest concern
to the international community (Hristov et al. 2011; Lenka et al. 2015). Ruminants
produce CO_2 and CH_4 in the upper part of the rumen stomach and release them into
the atmosphere through exhalation and belching. These gases are the most significant
GHGs. In the case of CH_4, the fermentation of ruminants produces c. 56% of total
GHG emissions (Broucek 2014), representing the largest quantity of *C-waste* content
emitted by the husbandry industry (Knapp et al. 2014; Ishler 2016) (Fig. 5.2). CO_2
emissions also occur, however, via urea hydrolysis in the manure pit. Hristov et al.
reported an estimate of c. 6,500.0 kg of CO_2 per dairy cow per year (Hristov et al.
2011).

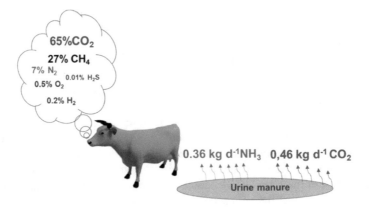

Fig. 5.2 Most representative emissions from a dairy cow

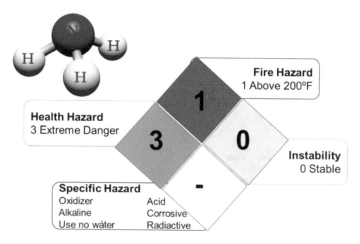

Fig. 5.3 Anhydrous ammonia hazard

The very harmful NH₃ is the main *N*-waste product emitted by the husbandry sector. As risks are involved in the emission of NH_3 (Fig. 5.3), this gas has a large ecological impact because it acts as the precursor of N_2O (Sutton et al. 2011). N_2O is a GHG molecule c. 300 times more harmful than CO_2 as a GHG. There are justifiable concerns about the fact that the dairy industrial division produces c. 38% of the total worldwide NH_3 emissions (Leytem and Dungan 2014; Leytem and Dungan 2014; USDA 2014).

In the range of c. 80% of the *N*-intake in cow diets is excreted in urine and faeces. NH_3 is generated from the action of the enzyme urease in the urea-urine hydrolysis process (Scheme 5.1) (Zhou et al. 2010; Reis et al. 2015; Mendes et al. 2017). This enzyme exists everywhere on farms, from the barn floors to the manure pit, which means urea hydrolysis is essentially an irreversible process (Barzagli et al. 2016). Volatilization of NH_3 is the most important pathway for loss of *N*-manure. A loss of up to 50% of the NH_4^+ content in manure was estimated by Leytem and Dungan (2014). NH_3 emissions depend on a number of factors directly linked with dairy farm activities (i.e. animal diet, farm temperature, farm wind features, farm manure management). In general terms, 13% of the emissions correspond to housing, 12% to manure management and 13% to land application of manure (Leytem and Dungan 2014; USDA 2014).

More than 50% of dairy cows in the world are in the top 10 countries. FAO reported more than 270 M (million) dairy cows in 2016 with an increase of 0.4%

$$(NH_2)_2CO_{(aq)} + H_2O_{(l)} \xrightarrow[\textit{urea hydrolysis}]{\textit{The global}} 2\,NH_{3(g)} + CO_{2(g)}$$

Scheme 5.1 Equation for global urea hydrolysis

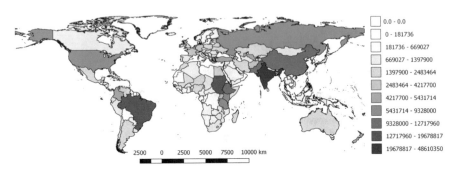

Fig. 5.4 Classification of countries by cow numbers (FAO 2012)

compared to 2015 (FAO 2012). India, Brazil, China and USA are the largest owners with a total close to 100 million dairy cows (Fig. 5.4). Countries such as China, the EU-28, the USA and Russia use intensive husbandry systems. They configured 'perfect barn' structures, which are characterised by the use of the best manure management strategies: (1) a slatted floor; (2) flushing the floor with water; (3) scraping the slatted floor, (4) separation of solid/liquid manure, and (5) an under-barn slurry manure pit (Reis et al. 2015; Baldini et al. 2016; Mendes et al. 2017). All these procedures were designed to reduce NH$_3$ emissions and guarantee animal welfare (Deng et al. 2015; Loyon et al. 2016; Aguirre-Villegas and Larson 2017).

5.2   CO$_2$ and NH$_3$ Estimated Emissions from Intensive Husbandry Production Systems

The most advanced dairy countries use intensive husbandry production in full confinement dairy cow farms. For CO$_2$ and NH$_3$ emissions estimates, the European Union (EU-28), China, the USA and Russia will be considered as an entity, henceforth called '*dairy entity*'. The '*dairy entity*' represents c. 53 M dairy cows (Alonso-Moreno et al. 2018).

5.2.1   Regarding CO$_2$ Emissions

A dairy cow exhales between 326.8 and 654.3 l of CO$_2$ a day. Dairy cows process carbohydrates as the main source of energy in the *Rumen Fermentation Processes* (RFP). The main by-products generated in the rumen stomach are listed in Fig. 5.2 (Knapp et al. 2014; Ishler 2016). Of them, CO$_2$ is the most abundant.

In the case of a 500 kg live weight animal, the estimate of CO$_2$ production per cow per day ranges from 326.8 to 654.3 l, and depends on the farm and the cow breed (Knapp et al. 2014). Using experimental values reported by Hristov et al, a

Table 5.1 Exhaled CO_2 emissions from different groupings of dairy cow

Dairy cow groupings	Exhaled CO_2 emissions		
Dairy cow	0.7 kg h^{-1}	17.8 kg d^{-1}	6500.0 kg yr^{-1}
Herd (140 cows)	103.6 kg h^{-1}	2493.1 kg d^{-1}	910.0 t yr^{-1}
Dairy EU-28 inventory	16.1 Mkg h^{-1}	409.4 Mkg d^{-1}	149.5 Mt yr^{-1}
Dairy China inventory	8.4 Mkg h^{-1}	213.6 Mkg d^{-1}	78.0 Mt yr^{-1}
Dairy US inventory	6.3 Mkg h^{-1}	160.2 Mkg d^{-1}	58.5 Mt yr^{-1}
Dairy Russia inventory	6.3 Mkg h^{-1}	160.2 Mkg d^{-1}	58.5 Mt yr^{-1}
Dairy entity	37.1 Mkg h^{-1}	943.4 Mkg d^{-1}	344.5 Mt yr^{-1}

dairy cow produces c. 6,500.0 kg of CO_2 a year (Hristov et al. 2011). In these terms, the '*dairy entity*' with c. 53.0 M dairy cows would exhale c. 344.5 Mt of CO_2 a year (Table 5.1).

The urine manure generated by a dairy cow produces *169.9* kg of CO_2 a year. CO_2 emissions occur by RFP and urea hydrolysis. According to Table 5.1, urea produces one mole of CO_2 by the microbial effect of the *enzyme urease* (Zhou et al. 2010; Aguilar 2012). Although urea content is only 4.7% of urine composition, a 500.0 kg live weight animal annually generates c. 169.9 kg of CO_2 by urea hydrolysis (Alonso-Moreno et al. 2018). Consequently, the '*dairy entity*' with c. 53 M dairy cows would produce c. 9.0 Mt of CO_2 per year.

The CO_2 concentration in a dairy cow stall is much higher than atmospheric CO_2 concentration. Mendes et al. published a specific emissions study of the vertical CO_2 concentration in a dairy cow stall. They reported CO_2 levels of 600 and 1000 ppm$_v$ (parts per million in volume) at 2 and 1 m above the barn floor, respectively (Mendes et al. 2015). These values are much higher than atmospheric CO_2 concentration (c. 400 ppm$_v$) (NASA 2017). Joo et al. 2015 supported this study with the measurement of CO_2 between 443 and 780 ppm$_v$ in a U.S. free-stall barn with natural ventilation.

The study reported by Mendes et al. enables the CO_2 concentration in a common dairy barn to be estimated. The estimate is carried out for a herd of 140 dairy cows confined in 2432 m^2, in a barn 64.0 m in length × 38.0 m in width × 4.0 m in height. If the wall were c. 2.0 m high, the barn volume would be c. 4864 m^3 (Mendes et al. 2015).

A dairy barn without an air-exchange system would be lethal for both workers and animals. Were the dairy barn not provided with good ventilation, the hypothetical CO_2 concentration, per the Mendes study (Mendes et al. 2015) would be c. 284,000.0 ppm$_v$ d^{-1}. The high density of the CO_2 would produce a CO_2-blanket close to the floor. Table 5.1 estimates a CO_2 emission of c. 0.46 kg cow^{-1} d^{-1} from urine manure, which means that the dairy herd (140 dairy cows) would produce c.64.4 kg d^{-1}; this means c. 40 times the CO_2 produced by urea hydrolysis. In other words, the CO_2 produced by the dairy herd in the barn is c. 2493.1 kg d^{-1}.

However, a natural ventilation system will allow the dilution of the CO$_2$ barn concentration from c. 284,000 to c. 700 ppm$_v$ d^{-1} (Mendes et al. 2015). This scenario would not allow capture of the CO$_2$.

To guarantee worker protection and reduce animal stress, forced ventilation systems (i.e. exhaust fans, tunnel ventilation or cross-over ventilation systems) are used to improve the air quality to an average CO$_2$ concentration in the barn of c. 700 ppm$_v$ (OSHA 2017). Additionally, these systems allow independence from wind speeds in the location of the dairy farm. This would be the appropriate scenario required to make CO$_2$ capture and subsequent transformation into a valuable chemical compound possible.

5.2.2   Regarding NH₃ Emissions

A dairy cow would generate c. 131.3 kg of NH$_3$ a year. The global NH$_{3(g)}$ emissions from dairy urine manure management can be estimated by the amount of liquid manure excreted by dairy cows annually (Hristov et al. 2011) (see Table 5.1 and Fig. 5.2 to better follow the discussion). Data collected by the 'The Merck Veterinary Manual' indicates values from 17 to 45 ml per kg of live weight animal (Fielder 2017). Considering each animal as a unit of 500.0 kg live weight, c. 5,860.9 kg per cow per year from urine manure. This value is very close to that reported in 'The Encyclopaedia of Soil Science data, 2nd Ed.' (Sejian et al. 2016). Considering an N-urea content of 4.7% in the urinary-urea manure, c. 231.7 kg of urea per cow per year is expected (Alonso-Moreno et al. 2018). Irreversible urea hydrolysis produces two moles of NH$_4^+$$_{(aq)}$ and one mole of CO$_{2(aq)}$ (Scheme 5.2) (Zhou et al. 2010; Aguilar 2012). Thus, a dairy cow would generate c. 131.3 kg of NH$_{3(aq)}$ in a year, in agreement with several previous studies (Nennich et al. 2005; Deng et al. 2015; Reis et al. 2015; Baldini et al. 2016).

Urea hydrolysys process by enzyme urease

Global Urea Hydrolysis Process

$$CO(NH_2)_{2(aq)} + H_2O_{(l)} \xrightarrow[\text{Urease}]{\text{Enzyme}} 2\ NH_4^+{}_{(aq)} + HCO_3^-{}_{(aq)} \longrightarrow 2\ NH_{3(g)} + CO_{2(g)} \quad (1)$$

C-waste speciation

$$CO_{2(g)} + H_2O_{(l)} \rightleftharpoons CO_{2(aq)} + H_2O_{(l)} \rightleftharpoons HCO_3^-{}_{(aq)} + H^+{}_{(aq)} \quad (2)$$

N-waste speciation

$$NH_{3(g)} + H_2O_{(l)} \rightleftharpoons NH_{3(aq)} + H_2O_{(l)} \rightleftharpoons NH_4^+{}_{(aq)} + OH^-{}_{(aq)} \quad (3)$$

Scheme 5.2 Global urea hydrolysis and speciation (*C-waste* and *N-waste*) by the urease enzyme

5.3 The *CO₂-RFP Strategy*

Chapter 4 described the best options for reducing the atmospheric concentration of CO_2. The use of ammonia as a scrubber is considered as one of the most suitable methods (Park et al. 2008; Spigarelli and Kawatra 2013). Even though RFP is the largest contributing source of GHG emissions in the world (Broucek 2014), there is currently no viable environmental action aimed at CO_2 reduction from RFP in the husbandry industry.

The '*CO₂-RFP Strategy*' is a novel, ground-breaking CDU environmental strategy in which *biogenic* CO_2 can be used to capture *biogenic* NH_3 in the husbandry industry (Alonso-Moreno et al. 2018). Based on a careful analysis of the literature on CO_2 absorption techniques used in the ternary chemical system CO_2-NH_3-H_2O, the '*CO₂-RFP Strategy*' is proposed (Alonso-Moreno et al. 2018), from a chemical point of view, to mitigate *biogenic* NH_3 generated in the manure pit of the barn. The strategy is intended to channel the exhaled CO_2 emissions from the dairy barn to the manure pit. This channelling would mean the in situ generation of a CO_2-blanket over the manure pit to control the NH_3 emissions (see illustration in Fig. 5.5) (Alonso-Moreno et al. 2018). The CO_2-blanket will then ensure the production of NH_4HCO_3 (see Fig. 5.5). NH_4HCO_3 is a suitable fertilizer, called ABC fertilizer (Li et al. 2003; Lee and Li 2003). According to Scheme 5.2, one extra mole of CO_2 would be required in the enclosure system. The vented flow of CO_2-enriched air will provide the extra CO_2 required, which will be added directly to the manure pit using pipes (OMAFRA 2016). The high solubility of CO_2 in ammonia solution, 200 times higher than in water, ensures the viability of the '*CO₂-RFP Strategy*' (Liu et al. 2009; Choi et al. 2012).

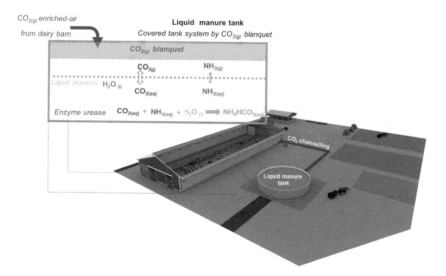

Fig. 5.5 The '*CO₂-RFP Strategy*' framework in a dairy farm

   The *'CO$_2$-RFP Strategy'* **would require the transformation of the manure pit into a closed system**. This procedure will be necessary to regulate the volatilization rate of CO$_2$ (Sejian et al. 2016) and to ensure the creation of a CO$_2$-blanket over the manure tank. A channel would also be required to connect the typical mechanical ventilation systems to the covered manure tank (OMAFRA 2016). Since OMAFRA recommends ventilation of c. 2550 m^3 h^{-1} on a dairy farm to treat heat stress in cows, the proposed *'CO$_2$-RFP Strategy'* would be plausible, as it only requires the capture of c. 125 m^3/h of the air in the barn. The channelling would allow the extra mole of CO$_2$ required in the reaction to be added to the liquid manure tank (see Fig. 5.5) (Zhou et al. 2010; Aguilar 2012).

5.3.1   The CO$_2$-RFP Strategy as a Business Model

The *'dairy entity'* comprises c. 53 M dairy cows and the dairy division of each country is made up of thousands of small dairy farms.

   The implementation of the *'CO$_2$-RFP Strategy'* would avoid c. 34.6 Mt of CO$_2$eq for the *NH$_3$* emission captured at the *'dairy entity'* per year. In a hypothetical implementation of the *'CO$_2$-RFP Strategy'* at the *'dairy entity'*, a total of c. 7.0 Mt of NH$_3$ emissions could be captured with c. 16.6 Mt of CO$_{2eq}$ emissions avoided per year. Moreover, c. 18.0 Mt of CO$_2$ emissions used in the synthesis of *biogenic* ABC fertilizer in a year should be added to the total of CO$_2$ stopped by the *'CO$_2$-RFP Strategy'*, which would yield c. 34.6 Mt of CO$_2$ avoided after implementation of the *strategy*.

   Table 5.2 shows an estimate of the NH$_3$ carbon footprint (CF) for the *'dairy entity'* if the *'CO$_2$-RFP Strategy'* was implemented. The CF of the NH$_3$ equivalent produced by synthetic methods is also included for comparison (YARA 2017).

   Implementation of the *'CO$_2$-RFP Strategy'* would yield c. 31.6 Mt of *biogenic*-ABC at the *'dairy entity'* in a year. This value is estimated using c. 7.0 Mt of *biogenic* NH$_3$ and c. 9.0 Mt of the *biogenic* CO$_2$ from the manure pit and the use of c. 9.0 Mt of *biogenic* CO$_2$ from the barn through the channelling process. Currently, total *N*-fertilizer production is estimated at c. 108.0 Mt. Of the large (YARA 2017) variety of *N*-fertilizer products, urea is the most important, with c. 58% of the total production (YARA 2017). China is the most important *N*-fertilizer country with an

Table 5.2 Estimate of NH$_3$ CF in the *'dairy entity'* after implementation of the *CO$_2$-RFP Strategy*

Dairy country	Dairy Herd (M cows)	NH$_3$ captured emissions (Mt)	CF per (t) of NH$_3$ produced	CO$_{2eq}$ (Mt) by NH$_3$ captured
EU-28	23.50	2.99	2.10	6.28
China	12.00	1.55	3.33	5.16
USA	9.00	1.16	2.10	2.43
Russia	9.00	1.16	2.10	2.43

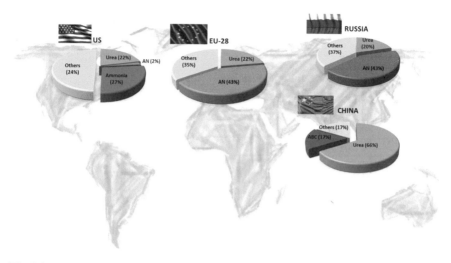

Fig. 5.6 *N*-fertilizer consumption in % at the '*dairy entity*' (YARA 2017)

annual production of c. 32.9 Mt. The EU-28 *N*-fertilizer consumption is 11.4 Mt in a proportion of 22% of urea and 43% of AN; China has a *N*-fertilizer consumption of 32.9 Mt, of which 66% is urea and 17% is ABC; from the 11.8 Mt of *N*-fertilizer consumed by USA, 22% is urea and 2% is AN; Russia had the lowest consumption of *N*-fertilizer with only 1.5 Mt, 20% of urea and 43% of AN (Fig. 5.6) (YARA 2017).

Regarding ABC production, China is the only manufacturer and consumer. China produced c. 50 Mt of ABC in 2003 by NH_3-based scrubbers from fossil fuel combustion (Li et al. 2003; Mani et al. 2006; Hsueh et al. 2010; Zhou et al. 2010; Zhuang et al. 2011; Zhao et al. 2012; Yu and Feron 2016). The implementation of the '*CO₂-RFP Strategy*' in China could increase its current ABC production to c. 7.2 Mt. Similarly, the implementation of the strategy at the '*dairy entity*' could produce enough *N*-fertilizer to replace the urea and AN consumption at the entity; 5.5 Mt of *biogenic*-ABC for Russia, 14.1 Mt for the EU-28 and 5.5 Mt for US. Furthermore, in the hypothetical case of implementation of the '*CO₂-RFP Strategy* at the '*dairy entity*', ABC production could substitute the most-used N-fertilizer in each country; 14.24 and 5.55 Mt of AN for EU-28 and Russia, and 2.70 and 2.10 Mt of urea for China and U.S. (see Table 5.3)

The '*CO₂-RFP Strategy*' would reduce the *N*-fertilizer CF produced by standard synthetic methods. Regarding the most important *N*-fertilizer production at the '*dairy entity*', the CF of urea is 5.33 t per tonne produced in China, 3.42 in the EU-28, 3.78 in the USA and 7.23 in Russia, whereas the CF of AN production is 3.4 t per tonne produced in the EU-28 and 7.2 in Russia. The NH_4^+-content of *biogenic* ABC produced by *CO₂-RFP strategy* at the 'diary entity' would be 14.24, 2.70, 2.10 and 5.5 Mt in EU-28, China, USA and Russia, respectively. Consequently, the estimated CO_2 equivalent avoided emission after implementation of the strategy would be 48.7, 14.40, 7.94 and 40.13 Mt (YARA 2017; Alonso-Moreno et al. 2018).

Table 5.3 Estimate of ABC production for the implementation of the *CO$_2$-RFP strategy* at the *'dairy entity''*. CO$_2$ captured emissions and CO$_2$ avoided emissions through the implementation of the *'CO$_2$-RFP strategy''*

Dairy country	ABC$_{eq}$ production (Mt)	CF per t N-Fertilizer equivalent	CO$_2$ (Mt) *captured emissions*	CO$_2$eq (Mt) *emissions*
EU-28	14.24 (AN)	3.42	7.81	48.70
Russia	5.55 (AN)	7.23	3.04	40.13
China	2.70 (Urea)	5.33	4.01	11.40
US	2.10 (Urea)	3.78	3.04	7.94

The hypothetical implementation of the CO$_2$-RFP strategy in the 'dairy entity' would generate nearly five times the classical ABC production. More importantly, the use of this new CDU proposal would give rise to c. 101.13 Mt of CO$_2$ equivalent avoided emissions. (Alonso-Moreno et al. 2018)

5.3.2   The CO$_2$-RFP Strategy *Regarding Negative Emissions*

Chapter 4 discussed the NETs concept and the most important technologies proposed to reduce the CO$_2$ atmospheric concentration. Anthropogenic activities carried out with reduction of CO$_2$ CF should be considered as NETs. *Biogenic*-ABC production through implementation of the *'CO$_2$-RFP strategy'* at the *'dairy entity'* can be analysed as CO$_2$eq *negative emissions*.

The *CO$_2$-RFP Strategy* **would generate c. 52.6 Mt of CO$_{2eq}$** *negative emissions* **annually using** *biogenic*-**NH$_3$**. The carbon tetrahedral diagram (CTD) in Fig. 5.7a illustrates this estimate. The movement of c. 36.0 Mt from the atmospheric pool to

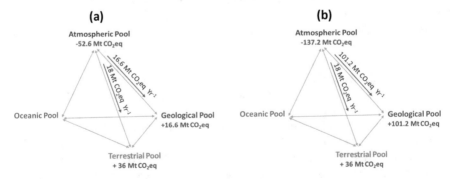

Fig. 5.7   a CTD regarding the use of *biogenic*-NH$_3$ by the implementation of the *'CO$_2$-RFP strategy'*. **b** CTD regarding the production of *biogenic*-ABC by the implementation of the *'CO$_2$-RFP Strategy'*

the terrestrial pool (c. 9.0 Mt of CO_2 in urea-content; c. 9.0 Mt of CO_2 capture in ABC-content and the 18.0 Mt of CO_2 avoided emissions) and the c. 16.6 Mt avoided emissions by substitution of the N-fertilizer classical synthesis and generated the above-mentioned c. 52.6 Mt of CO_2eq negative emissions (Alonso-Moreno et al. 2018). The urine-urea content excreted by the cows of the dairy entity produces c. 9.0 Mt of CO_2 and c.7.0 Mt of NH_3 in *C-waste* and *N-waste*, respectively. Total NH_3 emissions from the N-waste requires c. 18 Mt of CO_2. Additionally, the c. 7.0 Mt of NH_3 represents c. 16.6 Mt of CO_2eq avoided emissions due to the CF of industrial NH_3 synthesis. The geological pool would save c. 16.6 Mt of CO_2eq because fossil fuel combustion would not be necessary in NH_3 production.

The '*CO₂-RFP Strategy*' would generate c. 137.2 Mt of CO₂eq *negative emissions* annually, producing *biogenic*-ABC. *Biogenic*-ABC production through the '*CO₂-RFP strategy*' is equivalent to c. 24.54 Mt of traditional *N*-fertilizer crops obtained by synthetic methods. Thus, with the '*CO-RFP Strategy*' implemented at the '*dairy entity*', the EU-28 could replace by ABC c. 14.2 Mt of AN fertilizer; China and the USA c. 2.7 and 2.1 Mt of urea, respectively, and Russia c. 5.5 Mt of AN (see Table 5.3). This situation would avoid the emission of c. 101.2 Mt of CO_2 in *N*-fertilizer production at the '*dairy entity*' (Alonso-Moreno et al. 2018). The use of c. 18 Mt of biogenic CO_2 from the dairy farm would avoid the emission of c. 137.2 Mt of CO_2 (see CTD in Fig. 5.7b).

5.3.3 The CO₂-RFP Strategy *with Regard to Sustainability*

Any anthropogenic process must be a sustainable process (see Chap. 4).

The *CO₂-RFP Strategy* and *biogenic*-ABC production would preserve sustainability on a dairy farm. Figure 5.8 shows the *tetrahedral economy diagram* of the '*CO₂-RFP Strategy*' for *biogenic*-ABC production. The use of this diagram will

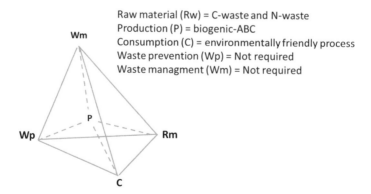

Raw material (Rw) = C-waste and N-waste
Production (P) = biogenic-ABC
Consumption (C) = environmentally friendly process
Waste prevention (Wp) = Not required
Waste managment (Wm) = Not required

Fig. 5.8 *Tetrahedral economy diagram.* The main parameter of a sustainable production process

clarify the connection between all the concepts to be considered in any anthropogenic process for it to be called environmentally sustainable.

Since the *CO$_2$-RFP Strategy* is a new production process, the sustainability of the process must be analysed. (1) *Raw materials*: both *waste-CO$_2$* and *waste-NH$_3$* livestock by-products from a dairy farm are the raw material, which implies that the production process renders a large amount of emissions into the atmosphere. (2) Consumption: N-fertilizer replaces traditional industrial methods for environmentally friendly processes. (3) Waste prevention: there is no extra by-product generated by the *CO$_2$-RFP Strategy*. (4) Waste management: there is no additional treatment required by the *CO$_2$-RFP Strategy*.

5.4   Conclusions

The *pros* and *cons* of the implementation of the '*CO$_2$-RFP Strategy*' must be analysed, in order to assess its viability. Pros: (1) significant improvement in the work conditions (lower CO$_2$ and NH$_3$ emissions), (2) guaranteed animal health at the farm (good temperature, low moisture and high-quality air in the barn), (3) low investment expenditure for the implementation of the environmental strategy (most of the technical facilities required already exist on a modern farm), (4) odour reduction (low NH$_3$ emissions), and (5) generation of sustainable and environmentally friendly dairy farms (*N*-fertilizer production from dairy farm raw materials). Cons: (1) highly confined husbandry system, and (2) dairy cows are considered as a milk production unit.

References

M.J. Aguilar, Urine as a CO$_2$ absorbent. J. Hazard. Mater. **213–214**, 502–504 (2012). https://doi.org/10.1016/j.jhazmat.2012.01.087

H.A. Aguirre-Villegas, R.A. Larson, Evaluating greenhouse gas emissions from dairy manure management practices using survey data and lifecycle tools. J. Clean. Prod. **143**, 169–179 (2017). https://doi.org/10.1016/j.jclepro.2016.12.133

C. Alonso-Moreno, S. García-Yuste, Environmental potential of the use of CO$_2$ from alcoholic fermentation processes. The CO$_2$-AFP strategy. Sci. Total Environ. **568**, 319–326 (2016). https://doi.org/10.1016/j.scitotenv.2016.05.220

C. Alonso-Moreno, J.J. Garde, J.E. Zafrilla et al., The Carbon Dioxide-rumen fermentation processes-strategy, a proposal to sustain environmentally friendly dairy farms. J. Clean. Prod. **204**, 735–743 (2018). https://doi.org/10.1016/j.jclepro.2018.08.295

K. Armstrong, P. Styring, Assessing the potential of utilization and storage strategies for post-combustion CO$_2$ emissions reduction. Front. Energy Res. **3** (2015). https://doi.org/10.3389/fenrg.2015.00008

C. Baldini, F. Borgonovo, D. Gardoni, M. Guarino, Comparison among NH$_3$ and GHGs emissive patterns from different housing solutions of dairy farms. Atmos. Environ. **141**, 60–66 (2016). https://doi.org/10.1016/j.atmosenv.2016.06.047

F. Barzagli, F. Mani, M. Peruzzini, Carbon dioxide uptake as ammonia and amine carbamates and their efficient conversion into urea and 1,3-disubstituted ureas. J. CO_2 Util. **13**, 81–89. (2016). https://doi.org/10.1016/j.jcou.2015.12.006

J. Broucek, Production of methane emissions from ruminant husbandry: a review. J. Environ. Prot. (Irvine, Calif) **5**, 1482–1493 (2014). https://doi.org/10.4236/jep.2014.515141

B.G. Choi, H.S. Park, G.H. Kim et al., Analysis of CO_2-NH_3 reaction dynamics in an aqueous phase by PCA and 2D IR cos. J. Ind. Eng. Chem. **18**, 98–104 (2012). https://doi.org/10.1016/j.jiec.2011.11.087

J. Deng, C. Li, Y. Wang, Modeling ammonia emissions from dairy production systems in the United States. Atmos. Environ. **114**, 8–18 (2015). https://doi.org/10.1016/j.atmosenv.2015.05.018

FAO, Statistics: Dairy cows. (2012)

Fielder, Urine volume and specific gravity—special subjects—merck veterinary manual. (2017). https://www.merckvetmanual.com/special-subjects/reference-guides/urine-volume-and-specific-gravity. Accessed 25 Nov 2019

A.N. Hristov, M. Hanigan, A. Cole et al., Review: Ammonia emissions from dairy farms and beef feedlots. Can. J. Anim. Sci. **91**, 1–35 (2011)

H.T. Hsueh, C.L. Hsiao, H. Chu, Removal of CO_2 from flue gas with ammonia solution in a packed tower. Sustain. Environ. Res. **20**, 1–7 (2010)

IEA, CO_2 emissions. (2017). https://www.iea.org/statistics/co2emissions/. Accessed 25 Nov 2019

IPCC, Climate change 2007 Synthesis report. Intergov. Panel Clim. Chang. (Core Writ Team IPCC 104). (2007). https://doi.org/10.1256/004316502320517344

V.A. Ishler, Carbon, Methane emissions and the dairy cow. (2016). https://extension.psu.edu/carbon-methane-emissions-and-the-dairy-cow. Accessed 25 Nov 2019

H.S. Joo, P.M. Ndegwa, A.J. Heber et al., Greenhouse gas emissions from naturally ventilated freestall dairy barns. Atmos. Environ. **102**, 384–392 (2015). https://doi.org/10.1016/j.atmosenv.2014.11.067

A.W. Kleij, M. North, A. Urakawa, CO_2 catalysis. Chemsuschem **10**, 1036–1038 (2017)

J.R. Knapp, G.L. Laur, P.A. Vadas et al., Invited review: Enteric methane in dairy cattle production: Quantifying the opportunities and impact of reducing emissions. J. Dairy Sci. **97**, 3231–3261 (2014)

J.W. Lee, R. Li, Integration of fossil energy systems with CO_2 sequestration through NH_4HCO_3 production. Energy Convers. Manag. **44**, 1535–1546 (2003). https://doi.org/10.1016/S0196-8904(02)00149-8

S. Lenka, N.K. Lenka, V. Sejian, M. Mohanty, Contribution of agriculture sector to climate change. Climate Change Impact on Livestock Adapt. Mitig. 37–48 (2015)

A.B. Leytem, R.S. Dungan, Livestock GRACEnet: A workgroup dedicated to evaluating and mitigating emissions from livestock production. J. Environ. Qual. **43**, 1101 (2014). https://doi.org/10.2134/jeq2014.06.0264

X. Li, E. Hagaman, C. Tsouris, J.W. Lee, Removal of carbon dioxide from flue gas by ammonia carbonation in the gas phase. Energy Fuels **17**, 69–74 (2003). https://doi.org/10.1021/ef020120n

J. Liu, S. Wang, B. Zhao et al., Absorption of carbon dioxide in aqueous ammonia. Energy Procedia 933–940 (2009)

L. Loyon, C.H. Burton, T. Misselbrook et al., Best available technology for European livestock farms: Availability, effectiveness and uptake. J. Environ. Manage. **166**, 1–11 (2016)

F. Mani, M. Peruzzini, P. Stoppioni, CO_2 absorption by aqueous NH_3 solutions: speciation of ammonium carbamate, bicarbonate and carbonate by a 13C NMR study. Green Chem. **8**, 995 (2006). https://doi.org/10.1039/b602051h

L.B. Mendes, N. Edouard, N.W.M. Ogink et al., Spatial variability of mixing ratios of ammonia and tracer gases in a naturally ventilated dairy cow barn. Biosyst. Eng. **129**, 360–369 (2015). https://doi.org/10.1016/j.biosystemseng.2014.11.011

L.B. Mendes, J.G. Pieters, D. Snoek et al., Reduction of ammonia emissions from dairy cattle cubicle houses via improved management- or design-based strategies: A modeling approach. Sci. Total Environ. **574**, 520–531 (2017). https://doi.org/10.1016/j.scitotenv.2016.09.079

NASA, The carbon cycle. (2017). https://earthobservatory.nasa.gov/features/CarbonCycle. Accessed 25 Nov 2019

T.D. Nennich, J.H. Harrison, L.M. VanWieringen et al., Prediction of manure and nutrient excretion from dairy cattle. J. Dairy Sci. **88**, 3721–3733 (2005). https://doi.org/10.3168/jds.S0022-0302(05)73058-7

I. Omae, Recent developments in carbon dioxide utilization for the production of organic chemicals. Coord. Chem. Rev. **256**, 1384–1405 (2012)

OMAFRA, Dairy housing—ventilation options for free stall barns. (2016). http://www.omafra.gov.on.ca/english/engineer/facts/15-017.htm. Accessed 25 Nov 2019

OSHA, OSHA occupational chemical database | occupational safety and health administration. (2017). https://www.osha.gov/chemicaldata/. Accessed 25 Nov 2019

H.S. Park, Y.M. Jung, J.K. You et al., Analysis of the CO$_2$ and NH3 reaction in an aqueous solution by 2D IR COS: Formation of bicarbonate and carbamate. J. Phys. Chem. A **112**, 6558–6562 (2008). https://doi.org/10.1021/jp800991d

S. Reis, C. Howard, M.A. Sutton (eds.), *Costs of Ammonia Abatement and the Climate Co-Benefits* (Springer, Netherlands, Dordrecht, 2015)

V. Sejian, L. Samal, M. Bagath et al., Manure management: gaseous emissions, 3rd Edn, in *Encyclopedia of Soil Science*. (2016). pp. 1400–1405

B.P. Spigarelli, S.K. Kawatra, Opportunities and challenges in carbon dioxide capture. J. CO$_2$ Util. **1**, 69–87 (2013)

M.A. Sutton, O. Oenema, J.W. Erisman et al., Too much of a good thing. Nature **472**, 159–161 (2011)

United Nations, The Paris agreement | UNFCCC. (2015). https://unfccc.int/process-and-meetings/the-paris-agreement/the-paris-agreement. Accessed 26 Nov 2019

USDA, USDA agricultural air quality task force | NRCS. (2014). https://www.nrcs.usda.gov/wps/portal/nrcs/main/national/air/taskforce/. Accessed 26 Nov 2019

USEPA, United States environmental protection agency | US EPA. (2017). https://www.epa.gov/. Accessed 26 Nov 2019

YARA, Yara International. (2017). https://www.yara.com/. Accessed 26 Nov 2019

H. Yu, P.H.M. Feron, Aqueous ammonia-based post-combustion CO$_2$ capture, in *Absorption-Based Post-Combustion Capture of Carbon Dioxide*. (2016). pp. 283–301

B. Zhao, Y. Su, W. Tao et al., Post-combustion CO$_2$ capture by aqueous ammonia: A state-of-the-art review. Int. J. Greenh. Gas Control **9**, 355–371 (2012)

W. Zhou, B. Zhu, Q. Li et al., CO$_2$ emissions and mitigation potential in China's ammonia industry. Energy Policy **38**, 3701–3709 (2010). https://doi.org/10.1016/j.enpol.2010.02.048

Q. Zhuang, R. Pomalis, L. Zheng, B. Clements, Ammonia-based carbon dioxide capture technology: Issues and solutions. Energy Procedia 1459–1470 (2011)